Praise for
Special Relativity and Classical Field Theory

"Thrilling...Susskind's approach is to subject the novice to a historical mathematical boot camp to make the path seem natural, and ultimately easier....At once witty and insightful."

—*Nature*

"Susskind is meticulous in explaining each step of applying the equations to the physical phenomena he describes, and he excels in explaining why the math works the way it does....If you're intrepid enough...you'll emerge with a much deeper appreciation for the true meaning of Einstein's relativity, Maxwell's equations, and many other aspects of fundamental physics."

—*Science News*

"Susskind and Friedman follow their collaboration on *Quantum Mechanics* by probing the mathematical nitty-gritty of field theory and Einstein's theory of special relativity in the third installment of the Theoretical Minimum series.... Enthusiastic discussion is seasoned with wry humor....Well paced...Clear and concise."

"Relativity and field theory are absolutely central to modern physics. Here they are explained masterfully, with insight and wit. This is physics the way it's really done, in all its glory, nothing swept under the rug."

"A true masterpiece that will stand the test of time. I only wish this book was around when I was learning classical physics. *Special Relativity and Classical Field Theory* was joyful to read, filled with insights and secrets that will prepare you for research."

This book is the third volume of the Theoretical Minimum series. The first volume, *The Theoretical Minimum: What You Need to Know to Start Doing Physics*, covered classical mechanics, which is the core of any physics education. We will refer to it from time to time simply as *Volume I*. The second book (*Volume II*) explains quantum mechanics and its relationship to classical mechanics. This third volume covers special relativity and classical field theory.

The books in this series run parallel to Leonard Susskind's videos, available on the Web through Stanford University (see www.theoreticalminimum.com for a listing). While covering the same general topics as the videos, the books contain additional details and topics that don't appear in the videos.

SPECIAL
RELATIVITY AND
CLASSICAL
FIELD THEORY

Also by the authors

Quantum Mechanics: The Theoretical Minimum
by Leonard Susskind and Art Friedman

*The Theoretical Minimum: What You Need to Know to Start
Doing Physics*
by Leonard Susskind and George Hrabovsky

*The Black Hole War: My Battle with Stephen Hawking
to Make the World Safe for Quantum Mechanics*
by Leonard Susskind

*The Cosmic Landscape:
String Theory and the Illusion of Intelligent Design*
by Leonard Susskind

*An Introduction to Black Holes, Information
and the String Theory Revolution: The Holographic Universe*
by Leonard Susskind and James Lindesay

SPECIAL RELATIVITY AND CLASSICAL FIELD THEORY

THE THEORETICAL MINIMUM

LEONARD SUSSKIND
AND ART FRIEDMAN

BASIC BOOKS
New York

Basic Books
Hachette Book Group
1290 Avenue of the Americas, New York, NY 10104
www.basicbooks.com

Printed in the United States of America

First Trade Paperback Edition: May 2019
Originally published in hardcover and ebook by Basic Books in September 2017

Published by Basic Books, an imprint of Perseus Books, LLC, a subsidiary of Hachette Book Group, Inc. The Basic Books name and logo is a trademark of the Hachette Book Group.

The Hachette Speakers Bureau provides a wide range of authors for speaking events. To find out more, go to www.hachettespeakersbureau.com or call (866) 376-6591.

The publisher is not responsible for websites (or their content) that are not owned by the publisher.

Library of Congress Control Number: 2017935228

ISBNs: 978-0-465-09334-2 (hardcover); 978-0-465-09335-9 (e-book); 978-1-5416-7406-6 (paperback)

LSC-C

10 9 8 7 6 5 4 3 2 1

Contents

To my father and my hero,

a man of courage,

Benjamin Susskind

—LS

To my wife, Maggie,

and her parents,

David and Barbara Sloan

—AF

Preface

This book is one of several that closely follow my Internet course series, The Theoretical Minimum. My coauthor, Art Friedman, was a student in these courses. The book benefited from the fact that Art was learning the subject and was therefore sensitive to the issues that might be confusing to the beginner. During the course of writing, we had a lot of fun, and we've tried to convey some of that spirit with a bit of humor. If you don't get it, ignore it.

The two previous books in this series cover classical mechanics and basic quantum mechanics. So far, we have not studied light, and that's because light is a relativistic phenomenon— a phenomenon that has to do with the special theory of relativity, or SR as we'll sometimes call it. That's our goal for this book: SR and classical field theory. Classical field theory means electromagnetic theory—waves, forces on charged particles, and so on—in the context of SR. Special relativity is where we'll begin.

Leonard Susskind

My parents, the children of immigrants, were bilingual. They taught us kids some Yiddish words and phrases but mainly reserved that language for themselves, often to say things they did not want us to understand. Many of their secret conversations were accompanied by loud peals of laughter.

Yiddish is an expressive language; it's well suited to great literature as well as to daily life and down-to-earth humor. It bothers me that my own comprehension is so limited. I'd love to read all the great works in the original, but frankly I'd be happy enough just to get the jokes.

A lot of us have similar feelings about mathematical physics. We want to understand the great ideas and problems and engage our own creativity. We know there's poetry to be read and written, and we're eager to participate in some fashion. All we lack is that "secret" language. In this series, our goal is to teach you the language of physics and show you some of the great ideas in their native habitat.

If you join us, you'll be able to wrap your head around a good portion of twentieth-century physics. You'll certainly be equipped to understand much of Einstein's early work. At a minimum, you'll "get the jokes" and the serious ideas that underlie them. To get you started, we've thrown in a few jokes of our own, including some real groaners.

I'm delighted to acknowledge everyone who helped and supported us along the way. It may be a cliché to say "we couldn't have done it without you," but it also happens to be true.

Working with the professionals at Brockman, Inc., and Basic Books is always a pleasure as well as a learning experience. John Brockman, Max Brockman, and Michael Healey played a critical role in transforming our idea into a real project. From there, TJ Kelleher, Hélène Barthélemy,

Carrie Napolitano, and Melissa Veronesi walked us through the editorial and production process with great skill and understanding. Laura Stickney of Penguin Books coordinated the publication of the UK edition so smoothly, we hardly saw it happening. Copyeditor Amy J. Schneider made substantial improvements to our initial manuscript, as did proofreaders Lor Gehret and Ben Tedoff.

A number of Leonard's former students generously offered to review the manuscript. This was no small task. Their insights and suggestions were invaluable, and the book is far better as a result. Our sincere thanks go to Jeremy Branscome, Byron Dom, Jeff Justice, Clinton Lewis, Johan Shamril Sosa, and Dawn Marcia Wilson. We are grateful to readers Beike Jia, Filip Van Lijsebetten, and Jo Ann Rees for their valuable corrections to the first printing.

As always, the warmth and support I've received from family and friends has seen me through this project. My wife, Maggie, spent hours creating and re-creating the two Hermann's Hideaway drawings, getting them done on time while dealing with the illness and passing of her mother.

This project has afforded me the luxury of pursuing two of my life passions at the same time: graduate level physics and fourth-grade humor. In this respect, Leonard and I are a perfect team, and collaborating with him is an unmitigated pleasure.

Art Friedman

Introduction

Dear readers and students of The Theoretical Minimum,

Hello there, and welcome back to Lenny & Art's Excellent Adventure. We last left the intrepid pair recovering from a wild rollicking roller coaster ride through the quantum world of entanglement and uncertainty. They were ready for something sedate, something reliable and deterministic, something classical. But the ride continues in *Volume III*, and it's no less wild. Contracting rods, time dilation, twin paradoxes, relative simultaneity, stretch limousines that do and don't fit into Volkswagen-size garages. Lenny and Art are hardly finished with their madcap adventure. And at the end of the ride Lenny tricks Art with a fake monopole.

Well, maybe that is a bit overwrought, but to the beginner the relativistic world is a strange and wondrous fun house, full of dangerous puzzles and slippery paradoxes. But we'll be there to hold your hand when the going gets tough. Some basic grounding in calculus and linear algebra should be good enough to get you through.

Our goal as always is to explain things in a completely serious way, without dumbing them down at all, but also without explaining more than is necessary to go to the next step. Depending on your preference, that could be either quantum field theory or general relativity.

It's been a while since Art and I published *Volume II* on quantum mechanics. We've been tremendously gratified by the thousands of e-mails expressing appreciation for our efforts, thus far, to distill the most important theoretical principles of physics into TTM.

The first volume on classical mechanics was mostly about the general framework for classical physics that was set up in the nineteenth century by Lagrange, Hamilton, Poisson, and other greats. That framework has lasted, and provides the underpinning for all modern physics, even as it grew into quantum mechanics.

Quantum mechanics percolated into physics starting from the year 1900, when Max Planck discovered the limits of classical physics, until 1926 when Paul Dirac synthesized the ideas of Planck, Einstein, Bohr, de Broglie, Schrödinger, Heisenberg, and Born into a consistent mathematical theory. That great synthesis (which, by the way, was based on Hamilton's and Poisson's framework for classical mechanics) was the subject of TTM *Volume II*.

In *Volume III* we take a historical step back to the nineteenth century to the origins of modern field theory. I'm not a historian, but I think I am accurate in tracing the idea of a field to Michael Faraday. Faraday's mathematics was rudimentary, but his powers of visualization were extraordinary and led him to the concepts of electromagnetic field, lines of force, and electromagnetic induction. In his intuitive way he understood most of what Maxwell later combined into his unified equations of electromagnetism. Faraday was lacking one element, namely that a changing electric field leads to effects similar to those of an electric current.

It was Maxwell who later discovered this so-called displacement current, sometime in the early 1860s, and then went on to construct the first true field theory: the theory of electromagnetism and electromagnetic radiation. But Maxwell's theory was not without its own troubling confusions.

The problem with Maxwell's theory was that it did not seem to be consistent with a basic principle, attributed to Galileo and clearly spelled out by Newton: All motion is relative. No (inertial) frame of reference is more entitled to be thought of as at rest than any other frame. However this principle was at odds with electromagnetic theory, which predicted that light moves with a distinct velocity $c = 3 \times 10^8$ meters per second. How could it be possible for light to have the same velocity in every frame of reference? How could it be that light travels with the same velocity in the rest frame of the train station, and also in the frame of the speeding train?

Maxwell and others knew about the clash, and resolved it the simplest way they knew how: by the expedient of tossing out Galileo's principle of relative motion. They pictured the world as being filled with a peculiar substance—the ether—which, like an ordinary material, would have a rest frame in which it was not moving. That's the only frame, according to the etherists, in which Maxwell's equations were correct. In any other frame, moving with respect to the ether, the equations had to be adjusted.

This was the status until 1887 when Albert Michelson and Edward Morley did their famous experiment, attempting to measure the small changes in the motion of light due to the motion of Earth through the ether. No doubt most readers know what happened; they failed to find any. People tried to explain away Michelson and Morley's result. The simplest idea was called ether drag, the idea being that the ether is dragged along with Earth so that the Michelson-Morley experiment was really at rest with respect to the ether. But no matter how you tried to rescue it, the ether theory was ugly and ungainly.

According to his own testimony, Einstein did not know about the Michelson-Morley experiment when in 1895 (at age sixteen), he began to think about the clash between electromagnetism and the relativity of motion. He simply felt intuitively that the clash somehow was not real. He based his thinking on two postulates that together seemed irreconcilable:

1. The laws of nature are the same in all frames of reference. Thus there can be no preferred ether-frame.
2. It is a law of nature that light moves with velocity c.

As uncomfortable as it probably seemed, the two principles together implied that light must move with the same velocity in all frames.

It took almost ten years, but by 1905 Einstein had reconciled the principles into what he called the special theory of relativity. It is interesting that the title of the 1905 paper did not contain the word *relativity* at all; it was "On the Electrodynamics of Moving Bodies." Gone from physics was the ever more complicated ether; in its place was a new theory of space and time. However, to this day you will still find a residue of the ether theory in textbooks, where you will find the symbol ϵ_0, the so-called dielectric constant of the vacuum, as if the vacuum were a substance with material properties. Students new to the subject often encounter a great deal of confusion originating from conventions and jargon that trace back to the ether theory. If I've done nothing else in these lectures, I tried to get rid of these confusions.

As in the other books in TTM I've kept the material to the minimum needed to move to the next step—depending on your preference, either quantum field theory or general relativity.

You've heard this before: Classical mechanics is intuitive; things move in predictable ways. An experienced ballplayer can take a quick look at a fly ball and from its location and its velocity know where to run in order to be there just in time to catch the ball. Of course, a sudden unexpected gust of wind might fool him, but that's only because he didn't take into account all the variables. There is an obvious reason why classical mechanics is intuitive: Humans, and animals before them, have been using it many times every day for survival.

In our quantum mechanics book, we explained in great detail why learning that subject requires us to *forget* our physical intuition and replace it with something entirely different. We had to learn new mathematical abstractions and a new way of connecting them to the physical world. But what about special relativity? While quantum mechanics explores the world of the VERY SMALL, special relativity takes us into the realm of the VERY FAST, and yes, it also forces us to bend our intuition. But here's the good news: The mathematics of special relativity is far less abstract, and we don't need brain surgery to connect those abstractions to the physical world. SR does stretch our intuition, but the stretch is far more gentle. In fact, SR is generally regarded as a branch of *classical* physics.

Special relativity requires us to rethink our notions of space, time, and especially simultaneity. Physicists did not make these revisions frivolously. As with any conceptual leap, SR was resisted by many. You could say that some physicists had to be dragged kicking and screaming to an acceptance of SR, and others never accepted it at all.[1] Why did most of them ultimately relent? Aside from the many

---

[1] Notably Albert Michelson, the first American to win a Nobel Prize in physics, and his collaborator Edward Morley. Their precise measurements provided strong confirmation of SR.

experiments that confirmed the predictions made by SR, there was strong *theoretical* support. The classical theory of electromagnetism, perfected by Maxwell and others during the nineteenth century, quietly proclaimed that "the speed of light is the speed of light." In other words, the speed of light is the same in every inertial (nonaccelerating) reference frame. While this conclusion was disturbing, it could not just be ignored—the theory of electromagnetism is far too successful to be brushed aside. In this book, we'll explore SR's deep connections to electromagnetic theory, as well as its many interesting predictions and paradoxes.

Lecture 1

The Lorentz Transformation

We open *Volume III* with Art and Lenny running for their lives.

Art: *Geez, Lenny, thank heavens we got out of Hilbert's place alive! I thought we'd never get disentangled. Can't we find a more classical place to hang out?*

Lenny: *Good idea, Art. I've had it with all that uncertainty. Let's head over to Hermann's Hideaway and see what's definitely happening.*

Art: *Where? Who is this guy Hermann?*

Lenny: *Minkowski? Oh, you'll love him. I guarantee there won't be any bras in Minkowski's space. No kets either.*

Lenny and Art soon wind up at Hermann's Hideaway, a tavern that caters to a fast-paced crowd.

Art: *Why did Hermann build his Hideaway way out here in the middle of—what? A cow pasture? A rice paddy?*

Lenny: *We just call it a field. You can grow just about*

anything you like; cows, rice, sour pickles, you name it. Hermann's an old friend, and I rent the land out to him at a very low price.

Art: *So you're a gentleman farmer! Who knew? By the way, how come everyone here is so skinny? Is the food that bad?*

Lenny: *The food is* great. *They're skinny because they're moving so fast. Hermann provides free jet packs to play with. Quick! Look out! Duck! DUCK!*

Art: *Goose! Let's try one out! We could both stand to get a bit thinner.*

More than anything else, the special theory of relativity is a theory about reference frames. If we say something about the physical world, does our statement remain true in a different reference frame? Is an observation made by a person standing still on the ground equally valid for a person flying in a jet? Are there quantities or statements that are invariant—that do not depend at all on the observer's reference frame? The answers to questions of this sort turn out to be interesting and surprising. In fact, they sparked a revolution in physics in the early years of the twentieth century.

1.1 Reference Frames

You already know something about reference frames. I talked about them in *Volume I* on classical mechanics. Cartesian coordinates, for example, are familiar to most people. A Cartesian frame has a set of spatial coordinates x, y, and z, and an origin. If you want to think concretely about what a coordinate system means, think of space as being filled up with a lattice of metersticks so that every point in space can be specified as being a certain number of meters to the left, a certain number of meters up, a certain number of meters in or out, from the origin. That's a coordinate system for space. It allows us to specify where an event happens.

In order to specify *when* something happens we also need a time coordinate. A reference frame is a coordinate system for both space and time. It consists of an x-, y-, z-, and t axis. We can extend our notion of concreteness by imagining that there's a clock at every point in space. We also imagine that we have made sure that all the clocks are synchronized, meaning that they all read $t = 0$ at the same instant, and that the clocks all run at the same rate. Thus a reference frame (or RF for simplicity) is a real or imagined lattice of metersticks together with a synchronized set of clocks at every point.

There are many ways to specify points in space and time, of course, which means we can have different RFs. We can translate the origin $x = y = z = t = 0$ to some other point, so that a location in space and time is measured relative to the new origin. We can also rotate the coordinates from one orientation to another. Finally, we consider frames that are moving relative to some particular frame. We can speak of *your* frame and *my* frame, and here we come to a key point: Besides the coordinate axes and the origin, a reference frame may be associated with an *observer* who can use all those clocks and metersticks to make measurements.

Let's assume you're sitting still at the center of the front row in the lecture hall. The lecture hall is filled with metersticks and clocks at rest in your frame. Every event that takes place in the room is assigned a position and a time by your sticks and clocks. I'm also in the lecture hall, but instead of standing still I move around. I might march past you moving to the left or to the right, and as I do I carry my lattice of clocks and metersticks with me. At every instant I'm at the center of my own space coordinates, and you are at the center of yours. Obviously my coordinates are different from yours. You specify an event by an x, y, z, and t, and I specify the *same* event by a different set of coordinates in order to account for the fact that I may be moving past you. In particular if I am moving along the x axis relative to you, we won't agree about our x coordinates. I'll always say that the end of my nose is at $x = 5$, meaning that it is five inches in front of the center of my head. However, you will say my nose is not at $x = 5$; you'll say my nose is moving, and that its position changes with time.

I might also scratch my nose at $t = 2$, by which I mean that the clock at the end of my nose indicated 2 seconds into the lecture when my nose was scratched. You might be tempted to think that your clock would also read $t = 2$ at the point where my nose was scratched. But that's exactly where relativistic physics departs from Newtonian physics. The assumption that all clocks in all frames of reference can be synchronized seems intuitively obvious, but it conflicts with Einstein's assumption of relative motion and the universality of the speed of light.

We'll soon elaborate on how, and to what extent, clocks at different places in different reference frames can be synchronized, but for now we'll just assume that at any given instant of time all of your clocks agree with each other, and they agree with my clocks. In other words we temporarily follow Newton and assume that the time coordinate is exactly the same for you as it is for me, and there's no ambiguity resulting from our relative motion.

1.2 Inertial Reference Frames

The laws of physics would be very hard to describe without coordinates to label the events that take place. As we've seen, there are many sets of coordinates and therefore many descriptions of the same events. What relativity meant, to Galileo and Newton as well as Einstein, is that the laws governing those events are the same in all inertial reference frames.[1] An inertial frame is one in which a particle, with no external forces acting on it, moves in a straight line with uniform velocity. It is obvious that not all frames are inertial. Suppose your frame is inertial so that a particle, thrown through the room, moves with uniform velocity when

---

[1] Sometimes we'll use the abbreviation IRF for *inertial reference frame.*

measured by your sticks and clocks. If I happen to be pacing back and forth, the particle will look to me like it accelerates every time I turn around. But if I walk with steady motion along a straight line, I too will see the particle with uniform velocity. What we may say in general is that any two frames that are both inertial must move with uniform relative motion along a straight line.

It's a feature of Newtonian mechanics that the laws of physics, $F = ma$ together with Newton's law of gravitational attraction, are the same in every IRF. I like to describe it this way: Suppose that I am an accomplished juggler. I have learned some rules for successful juggling, such as the following: If I throw a ball vertically upward it will fall back to the same point where it started. In fact, I learned my rules while standing on the platform of a railway station waiting for the train.

When the train stops at the station I jump on and immediately start to juggle. But as the train pulls out of the station, the old laws don't work. For a time the balls seem to move in odd ways, falling where I don't expect them. However, once the train gets going with uniform velocity, the laws start working again. If I'm in a moving IRF and everything is sealed so that I can't see outside, I cannot tell that I'm moving. If I try to find out by doing some juggling, I'll find out that my standard laws of juggling work. I might assume that I'm at rest, but that's not correct; all I can really say is that I'm in an inertial reference frame.

The principle of relativity states that *the laws of physics are the same in every IRF*. That principle was not invented by Einstein; it existed before him and is usually attributed to Galileo. Newton certainly would have recognized it. What new ingredient did Einstein add? He added one law of physics: the law that the speed of light is the speed of light, c. In units of meters per second, the speed of light is approximately 3×10^8. In miles per second it is about 186,000, and in light-years per year it is exactly 1. But once the units are chosen, Einstein's new law states that the velocity of light is the same for every observer.

When you combine these two ideas—that the laws of physics are the same in every IRF, and that it's a law of physics that light moves at a fixed velocity—you come to the conclusion that light must move with the same velocity in *every* IRF. That conclusion is truly puzzling. It led some physicists to reject SR altogether. In the next section, we'll follow Einstein's logic and find out the ramifications of this new law.

1.2.1   Newtonian (Pre-SR) Frames

In this section I will explain how Newton would have described the relation between reference frames, and the conclusions he would have made about the motion of light rays. Newton's basic postulate would have been that there exists a universal time, the same in all reference frames.

Let's begin by ignoring the y and z directions and focus entirely on the x direction. We'll pretend that the world is one-dimensional and that all observers are free to move along the x axis but are frozen in the other two directions of space. Fig. 1.1 follows the standard convention in which the x axis points to the right, and the t axis points up. These axes describe the metersticks and clocks in *your* frame—the frame at rest in the lecture hall. (I will arbitrarily refer to your frame as the *rest frame* and my frame as the *moving frame*.) We'll assume that in your frame light moves with its standard speed c. A diagram of this kind is called a *spacetime diagram*. You can think of it as a map of the world, but a map that shows all possible places *and* all possible times. If a

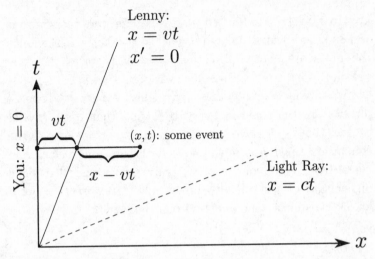

Figure 1.1: Newtonian Frames.

light ray is sent out from the origin, moving toward the right, it will move with a trajectory given by the equation

$$x = ct.$$

Similarly a light ray moving to the left would be represented by

$$x = -ct.$$

A negative velocity just means moving to the left. In the various figures that follow I will draw them as if my frame is moving to the right (v positive). As an exercise you can redraw them for negative v.

In Fig. 1.1, the light ray is shown as a dashed line. If the units for the axes are meters and seconds, the light ray will appear almost horizontal; it will move 3×10^8 meters to the right while moving vertically by only 1 second! But the numerical value of c depends entirely on the units we choose. Therefore, it is convenient to use some other units for the speed of light—units in which we see more clearly that the slope of the light-ray trajectory is finite.

Now let's add *my* frame, moving along the x axis relative to your frame, with a uniform velocity v.[2] The velocity could be either positive (in which case I would be moving to your right), negative (in which case I would be moving to your left), or zero (in which case we are at rest relative to each other, and my trajectory would be a vertical line in the figure).

I'll call the coordinates in my frame x' and t' instead of x and t. The fact that I am moving relative to you with a constant velocity implies that my trajectory through space-time is a straight line. You may describe my motion by the equation

$$x = vt$$

or

$$x - vt = 0,$$

where v is my velocity relative to you, as shown in Fig. 1.1. How do I describe my own motion? That's easy; I am always at the origin of my own coordinate system. In other words, I describe myself by the equation $x' = 0$. The interesting question is, how do we translate from one frame to the other; in other words, what's the relationship between your coordinates and mine? According to Newton, that relation would be

$$t' = t \tag{1.1}$$

$$x' = x - vt. \tag{1.2}$$

The first of these equations is Newton's assumption of

---

[2] We could also describe this as "my trajectory in your frame."

universal time, the same for all observers. The second just shows that my coordinate x' is displaced from your coordinate by our relative velocity times the time, measured from the origin. From this we see that the equations

$$x - vt = 0$$

and

$$x' = 0$$

have the same meaning. Eqs. 1.1 and 1.2 comprise the Newtonian transformation of coordinates between two inertial reference frames. If you know when and where an event happens in your coordinates, you can tell me in my coordinates when and where it happens. Can we invert the relationship? That's easy and I will leave it to you. The result is

$$t = t' \tag{1.3}$$

$$x = x' + vt'. \tag{1.4}$$

Now let's look at the light ray in Fig. 1.1. According to assumption, it moves along the path $x = ct$ in your frame. How do I describe its motion in my frame? I simply substitute the values of x and t from Eqs. 1.3 and 1.4 into the equation $x = ct$, to get

$$x' + vt' = ct',$$

which we rewrite in the form

$$x' = (c - v)t'.$$

Not surprisingly, this shows the light ray moving with velocity $(c - v)$ in my frame. That spells trouble for Einstein's new law—the law that all light rays move with the same speed c in every IRF. If Einstein is correct, then something is seriously wrong. Einstein and Newton cannot both be right: The speed of light cannot be universal if there is a universal time that all observers agree on.

Before moving on, let's just see what would happen to a light ray moving to the *left*. In your frame such a light ray would have the equation

$$x = -ct.$$

It's easy to see that in my frame the Newtonian rules would give

$$x' = -(c + v)t.$$

In other words, if I'm moving to your right, a light ray moving in the same direction travels a little slower (with speed $c - v$), and a light ray moving in the opposite direction travels a little faster (with speed $c + v$) relative to me. That's what Newton and Galileo would have said. That's what *everyone* would have said until the end of the nineteenth century when people started measuring the speed of light with great precision and found out that it's always the same, no matter how the inertial observers move.

The only way to reconcile this conflict is to recognize that something is wrong with Newton's transformation law between coordinates in different frames.[3] We need to figure out how to repair Eqs. 1.1 and 1.2 so that the speed of light is the same for both of us.

---

[3] This statement may sound glib. The fact is that many of the world's most talented physicists tried to make things work out without giving up on the equation $t' = t$. They all failed.

1.2.2   SR Frames

Before deriving our new transformation equations, let's revisit one of Newton's key assumptions. The assumption that is most at risk, and in fact the one that's wrong, is that simultaneity means the same thing in every frame—that if we begin with our clocks synchronized, and then I start to move, my clocks will remain synchronized with your clocks. We're about to see that the equation

$$t' = t$$

is *not* the correct relationship between moving clocks and stationary clocks. The whole idea of simultaneity is frame-dependent.

Synchronizing Our Clocks

Here's what I want you to imagine. We're in a lecture hall. You, a student, are sitting in the front row, which is filled with eager attentive students, and each student in the front row has a clock. The clocks are all identical and completely reliable. You inspect these clocks carefully, and make sure they all read the same time and tick at the same rate. I have an equivalent collection of clocks in my frame that are spread out relative to me in the same way as your clocks. Each of your clocks has a counterpart in my setup, and vice versa. I've made sure that my clocks are synchronized with each other and also with your clocks. Then I, together with all my clocks, start moving relative to you and your clocks. As each of my clocks passes each of yours, we check each other's clocks to see if they still read the same time and if not, how far out of whack each clock is compared to its counterpart. The answer may depend on each clock's position along the line.

Of course, we could ask similar questions about our metersticks, such as, "As I pass you, does my meterstick measure 1 in your coordinates?" This is where Einstein made his great leap. He realized that we have to be much more careful about how we define lengths, times, and simultaneity. We need to think *experimentally* about how to synchronize two clocks. But the one anchor that he held on to is the postulate that the speed of light is the same in every IRF. For that, he had to give up Newton's postulate of a universal time. Instead he found that "simultaneity is relative." We will follow his logic.

What exactly do we mean when we say that two clocks— let's call them A and B—are synchronized? If the two clocks are at the same location, moving with the same velocity, it should be easy to compare them and see if they read the same value of time. But even if A and B are standing still, say in your frame, but are not at the same position, checking if they are synchronized requires some thought. The problem is that light takes time to travel between A and B.

Einstein's strategy was to imagine a third clock, C, located midway between A and B.[4] To be specific, let's imagine all three clocks being located in the front row of the lecture hall. Clock A is held by the student at the left end of the row, clock B is held by the student at the right end, and clock C is at the center of the row. Great care has been taken to make sure that the distance from A to C is the same as the distance from B to C.

At exactly the time when the A clock reads noon it activates a flash of light toward C. Similarly when B reads noon it also sends a flash of light to C. Of course, both flashes will take some time to reach C, but since the velocity of light is the same for both flashes, and the distance they have to

_____

[4] This is actually a slight variation of Einstein's approach.

travel is the same, they both take the same time to get to C. What we mean by saying A and B are synchronized is that the two flashes will arrive at C at exactly the same time. Of course, if they don't arrive simultaneously, student C will conclude that A and B were not synchronized. She may then send a message to either A or B with instructions for how much to change their settings to get synchronized.

Suppose clocks A and B *are* synchronized in your frame. What happens in my moving frame? Let's say I'm moving to the right, and I happen to reach the midpoint C just as these two flashes are emitted. But the light doesn't get to C at noon; it gets there slightly later. By that time, I've already moved a little to the right of center. Since I'm right of center, the light ray coming from the left will reach me a little later than the light ray coming from the right. Therefore, I will conclude that your clocks are *not* synchronized, because the two light flashes reach me at two different times.

Evidently what you and I call synchronous—occurring at the same time—is not the same. Two events that take place at the same time in your frame take place at different times in my frame. Or at least that's what Einstein's two postulates force us to accept.

Units and Dimensions: A Quick Detour

Before moving ahead, we should pause briefly to explain that we'll be using two systems of units. Each system is well suited to its purpose, and it's fairly easy to switch from one system to the other.

The first system uses familiar units such as meters, seconds, and so on. We'll call them *common* or *conventional* units. These units are excellent for describing the ordinary world in which most velocities are far smaller than the speed of light. A velocity of 1 in those units means 1 meter per second, orders of magnitude less than c.

The second system is based on the speed of light. In this system, units of length and time are defined in a way that gives the speed of light a dimensionless value of 1. We call them *relativistic* units. Relativistic units make it easier to carry out derivations and notice the symmetries in our equations. We've already seen that conventional units are impractical for spacetime diagrams. Relativistic units work beautifully for this purpose.

In relativistic units, not only does c have a value of 1, but *all* velocities are dimensionless. For this to work out, we have to choose appropriately defined units of length and time—after all, velocity is a length divided by a time. If our time units are seconds, we choose light-seconds as our length units. How big is a light-second? We know that it's $186,000$ miles, but for our purposes that's unimportant. Here's what matters: A light-second is a unit of length, and by definition light travels 1 light-second per second! In effect, we're measuring both time *and* length in units of seconds. That's how velocity—a length divided by a time—gets to be dimensionless. When we use relativistic units, a velocity variable such as v is a dimensionless fraction of the speed of light. That's consistent with c itself having a value of 1.

In a spacetime diagram such as Fig. 1.2, the x and t axes are both calibrated in seconds.[5] The trajectory of a light ray makes equal angles with the x axis and with the t axis. Conversely, any trajectory that makes equal angles with the two axes represents a light ray. In your stationary RF, that angle is 45 degrees.

---

[5] You can think of the t axis being calibrated in light-seconds instead of seconds if you prefer, but it amounts to the same thing.

Knowing how to switch easily between the two types of units will pay off. The guiding principle is that mathematical expressions need to be dimensionally consistent in whatever system of units we're using at the time. The most common and useful trick in going from relativistic to conventional units is to replace v with v/c. There are other patterns as well, which typically involve multiplying or dividing by some appropriate power of c, the speed of light. We'll show examples as we go, and you'll find that these conversions are fairly simple.

Setting Up Our Coordinates—Again!

Let's go back to our two coordinate systems. This time, we'll be very careful about the exact meaning of the word *synchronous* in the moving RF. In the stationary RF, two points are synchronous (or simultaneous) if they're both on the same horizontal level in a spacetime diagram. The two points both have the same t coordinate, and a line connecting them is parallel to the x axis. That much Newton would have agreed with.

But what about the moving frame? We'll find out in a minute that in the *moving frame*, the point

$$x = 0, \ t = 0$$

is *not* synchronous with the other points on the x axis, but with an entirely different set of points. In fact, the whole surface that the moving frame calls "synchronous" is someplace else. How can we map out this surface? We'll use the synchronization procedure, described in a previous subsection (Synchronizing Our Clocks) and further illustrated in Fig. 1.2.

Drawing a spacetime diagram is usually the best way to understand a problem in relativity. The picture is always the same; x is the horizontal axis, and t is vertical. These coordinates represent a RF that you can think of as stationary. In other words, they represent *your* frame. A line that represents the trajectory of an observer moving through spacetime is called a *world line*.

With our axes in place, the next things to draw are the light rays. In Fig. 1.2, these are represented by the lines labeled $x = ct$ and $x = -ct$. The dashed line from point a to point b in the figure is also a light ray.

Back to the Main Road

Getting back to Fig. 1.2, let's sketch in an observer, Art, who's sitting in a railroad car moving to the right with constant speed v. His world line is labeled with equations that describe his motion. Once again, Art's frame will be moving such that $x' = x - vt$, exactly like the moving observer in Fig. 1.1.

Now let's figure out how to draw Art's x' axis. We begin by adding two more observers, Maggie and Lenny. Maggie is sitting in the rail car directly in front of Art (to your right), and Lenny's rail car is directly in front of Maggie's. Adjacent observers are separated from each other by one unit of length as measured in your frame (the rest frame). Equations for Maggie's and Lenny's world lines are shown in the figure. Because Maggie is located one unit to the right of Art, her trajectory is just $x = vt + 1$. Likewise, Lenny's trajectory is $x = vt + 2$. Art, Maggie, and Lenny are in the same moving frame. They're at rest with respect to each other.

Our first observer, Art, has a clock, and his clock happens to read 12 noon just as he arrives at the origin. We'll assume the clock in the rest frame also reads 12 noon at this event. We both agree to call 12 noon our "time zero," and we label our common origin with $(x = 0, t = 0)$ in your coordinates,

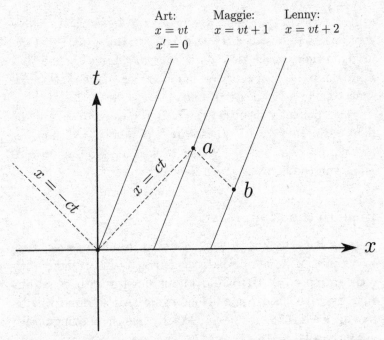

Figure 1.2: SR Frames Using Relativistic Units ($c = 1$). The equations associated with Art are two different ways to characterize his world line. The dashed lines are the world lines of light rays. The constants 1 and 2 in the equations for Maggie's and Lenny's world lines are not pure numbers. Their relativistic units are *seconds*.

and ($x' = 0, t' = 0$) in Art's coordinates. The moving observer and the stationary observer agree, by assumption, on the meaning of $t = 0$. For you (the stationary observer), t equals zero all along the horizontal axis. In fact, that's the *definition* of the horizontal axis: it's the line where all times for the stationary observer are zero.

Suppose Art sends out a light signal to Maggie from the origin. At some point—we don't know yet what that point is—Lenny will *also* send out a light signal toward Maggie.

He'll somehow arrange to do this in a way that both signals reach Maggie at the same instant. If Art's light signal starts at the origin, Maggie will receive it at point a in Fig. 1.2. From what point must Lenny send his light signal, if it is to reach at Maggie at the same time? We can find out by working backward. The world line of any light signal that Lenny sends to Maggie must make a 45-degree angle to the x axis. So all we need to do is construct a line at point a that slopes 45 degrees downward toward the right, and extend it until it crosses Lenny's path. This is point b in the figure. As we can easily see from the figure, point b lies above the x axis and not on it.

What we've just shown is that the origin and point b are *simultaneous* events in Art's frame! In other words, the moving observer (Art) will say that $t' = 0$ at point b. Why? Because in the moving frame of reference, Art and Lenny, who are equidistant from the central observer Maggie, sent her light signals that arrived at the same instant of time. So Maggie will say, "You guys sent me light signals at exactly the same instant of time, because they both arrived here at the same instant of time, and I happen to know that you're at equal distances from me."

Finding the x' Axis

We've established that Art's (and Maggie's and Lenny's) x axis is a line that joins the common origin (i.e., common to both frames) to point b. Our next task is to find out exactly where point b is. Once we figure out point b's coordinates, we'll know how to specify the direction of Art's x' axis. We'll go through this exercise in detail. It's a little cumbersome, but quite easy. There are two steps involved, the first being to find the coordinates of point a.

Point a sits at the intersection of two lines; the rightward moving light ray $x = ct$, and the line $x = vt + 1$, which is Maggie's world line. To find the intersection, we just substitute one equation into the other. Because we're using relativistic units for which the speed of light c is equal to 1, we can write the equation

$$x = ct$$

in an even simpler way,

$$x = t. \tag{1.5}$$

Substituting Eq. 1.5 into Maggie's world line,

$$x = vt + 1,$$

gives

$$t = vt + 1$$

or

$$t(1 - v) = 1.$$

Or, even better,

$$t_a = 1/(1 - v). \tag{1.6}$$

Now that we know the time coordinate of point a we can find its x coordinate. This is easily done by noticing that all along the light ray, $x = t$. In other words, we can just replace t with x in Eq. 1.6, and write

$$x_a = 1/(1 - v).$$

Voilà!—we've found point a.

With the coordinates of point a in hand, let's look at line ab. Once we have an equation for line ab, we can figure out where it intersects Lenny's world line, $x = vt + 2$. It takes a few steps, but they're fun and I don't know of any shortcuts.

Every line that slopes at 45 degrees, pointing downward to the right, has the property that $x + t$ is constant along that line. Every line sloping *upward* to the right at 45 degrees has the property that $x - t$ is constant. Let's take the line ab. Its equation is

$$x + t = some\ constant.$$

What is the constant? An easy way to find out is to take one point along the line and plug in specific values x and t. In particular, we happen to know that at point a,

$$x_a + t_a = 2/(1 - v).$$

Therefore, we know that this is true all along line ab, and the equation for that line must be

$$x + t = 2/(1 - v). \tag{1.7}$$

Now we can find b's coordinates by solving the simultaneous equations for line ab and for Lenny's world line. Lenny's world line is $x = vt + 2$, which we rewrite as $x - vt = 2$. Our simultaneous equations are

$$x + t = 2/(1 - v)$$

and

$$x - vt = 2.$$

The solution, after a bit of easy algebra, is

$$t_b = \frac{2v}{(1 - v^2)}$$

$$x_b = \frac{2}{(1 - v^2)}. \tag{1.8}$$

First, the most important point: t_b is not zero. Therefore point b, which is simultaneous with the origin in the moving frame, is *not* simultaneous with the origin in the rest frame.

Next, consider the straight line connecting the origin with point b. By definition, the slope of that line is t_b/x_b, and using Eqs. 1.8 we see that the slope is v. This line is nothing but the x' axis, and it is given, very simply, by the equation

$$t = vx. \tag{1.9}$$

Keep in mind as we go along that the velocity v can be positive or negative depending on whether I am moving to the right or to the left relative to your frame. For negative velocity you will have to redraw diagrams or just flip them horizontally.

Fig. 1.3 shows our spacetime diagram with the x' and t' axes drawn in. The line $t = vx$ (or what is really a three-dimensional surface in the spacetime map when the two other coordinates y, z are accounted for) has the important property that on it, all the clocks in the moving frame record the same value of t'. To give it a name, it is a *surface of simultaneity* in the moving frame. It plays the same role as the surface $t = 0$ does for the rest frame.

So far in this section, we've worked in relativistic units where the speed of light is $c = 1$. Here is a good opportunity to practice your skills in dimensional analysis and figure out what Eq. 1.9 would look like in conventional units of meters and seconds. In those units, Eq. 1.9 is not dimensionally consistent; the left side has units *seconds* and the right side has units *meters squared per second*. To restore consistency, we have to multiply the right side by an appropriate power of c. The correct factor is $1/c^2$:

$$t = \left(\frac{v}{c^2}\right)x. \tag{1.10}$$

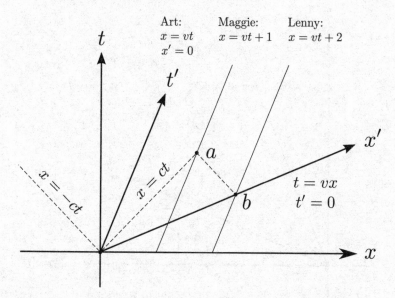

Figure 1.3: SR Frames with x' and t' Axes Shown.

The interesting thing about Eq. 1.10 is that it describes a straight line with the incredibly tiny slope $\dfrac{v}{c^2}$. For example, if v were 300 meters per second (roughly the speed of a jetliner) the slope would be $v/c^2 = 3 \times 10^{-15}$. In other words, the x' axis in Fig. 1.3 would be almost exactly horizontal. The surfaces of simultaneity in the rest and moving frames would almost exactly coincide just as they would in Newtonian physics.

This is an example of the fact that Einstein's description of spacetime reduces to Newton's if the relative velocity of the reference frames is much less than the speed of light. This, of course, is an important "sanity" check.

Now we can return to relativistic units with $c = 1$. Let's simplify our diagram and keep only the features we'll need going forward. The dashed line in Fig. 1.4 represents a light

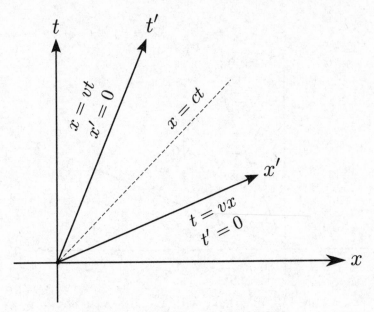

Figure 1.4: SR Frames Simplified.

ray, whose world line makes a 45-degree angle with both the t and x axes. Art's world line is shown as the t' axis. His x' axis is also labeled. Both of Art's axes are also labeled with appropriate equations. Notice the symmetry of his two axes: $x = vt$ and $t = vx$. These two lines are reflections of each other about the dashed light trajectory. They're related by interchanging t and x. Another way to say it is that they each make the same angle with their nearest unprimed axis—the x axis in the case of $t = vx$, and the t axis in the case of $x = vt$.

We've discovered two interesting things. First, if the speed of light really is the same in every frame, and you use light rays to synchronize clocks, then the pairs of events that are synchronous in one frame are not the same pairs that are synchronous in the other frame. Second, we've found what synchronicity actually means in the moving frame. It corresponds to surfaces that are *not* horizontal, but are tilted with slope v. We have figured out the directions of the x' and t' axes in Art's moving frame. Later on, we'll figure out how to mark off the intervals along these axes.

Spacetime

Let's pause for a moment to contemplate what we've found about space and time. Newton, of course, knew about both space and time but regarded them as entirely separate. To Newton, three-dimensional space was space, and time was universal time. They were entirely separate, the difference being absolute.

But maps like Figs. 1.3 and 1.4 indicate something that Newton could not have known, namely that in going from one inertial reference frame to another, the space and time coordinates get mixed up with each other. For example, in Fig. 1.3 the interval between the origin and point b represents two points at the same time in the moving frame. But in the rest frame, point b is not only shifted in space from the origin; it also is shifted in time.

Three years after Einstein's mighty 1905 paper that introduced the special theory of relativity, Minkowski completed the revolution. In an address to the 80th Assembly of German Natural Scientists and Physicians, he said,

> Space by itself, and time by itself, are doomed to fade away into mere shadows, and only a kind of union of the two will preserve an independent reality.

That union is four-dimensional, with coordinates t, x, y, z. Depending on our mood, we physicists sometimes call that union of space and time *spacetime*. Sometimes we call it

Minkowski space. Minkowski had another name for it. He called it *the world*.

Minkowski called the points of spacetime *events*. An event is labeled by the four coordinates t, x, y, z. By calling a point of spacetime an event, Minkowski did not mean to imply that something actually took place at t, x, y, z. Only that something *could* take place. He called the lines or curves describing trajectories of objects *world lines*. For example, the line $x' = 0$ in Fig. 1.3 is Art's world line.

This change of perspective from space and time to spacetime was radical in 1908, but today spacetime diagrams are as familiar to physicists as the palms of their hands.

Lorentz Transformations

An event, in other words a point of spacetime, can be labeled by the values of its coordinates in the rest frame or by its coordinates in the moving frame. We are talking about two different descriptions of a single event. The obvious question is, how do we go from one description to the other? In other words, what is the *coordinate transformation* relating the rest frame coordinates t, x, y, z to the coordinates t', x', y', z' of the moving frame?

One of Einstein's assumptions was that spacetime is everywhere the same, in the same sense that an infinite plane is everywhere the same. The sameness of spacetime is a symmetry that says no event is different from any other event, and one can choose the origin anywhere without the equations of physics changing. It has a mathematical implication for the nature of transformations from one frame to another. For example, the Newtonian equation

$$x' = x - vt \qquad (1.11)$$

is linear; it contains only first powers of the coordinates. Equation 1.11 will not survive in its simple form, but it does get one thing right, namely that $x' = 0$ whenever $x = vt$. In fact there is only one way to modify Eq. 1.11 and still retain its linear property, along with the fact that $x' = 0$ is the same as $x = vt$. It is to multiply the right side by a function of the velocity:

$$x' = (x - vt)f(v). \tag{1.12}$$

At the moment the function $f(v)$ could be any function, but Einstein had one more trick up his sleeve: another symmetry—the symmetry between left and right. To put it another way, nothing in physics requires movement to the right to be represented by positive velocity and movement to the left by negative velocity. That symmetry implies that $f(v)$ must not depend on whether v is positive or negative. There is a simple way to write any function that is the same for positive and negative v. The trick is to write it as a function of the square of the velocity v^2.[6] Thus, instead of Eq. 1.12, Einstein wrote

$$x' = (x - vt)f(v^2). \tag{1.13}$$

To summarize, writing $f(v^2)$ instead of $f(v)$ emphasizes the point that there is no preferred direction in space.

What about t'? We'll reason the same way here as we did for x'. We know that $t' = 0$ whenever $t = vx$. In other words, we can just invert the roles of x and t, and write

$$t' = (t - vx)g(v^2), \tag{1.14}$$

where $g(v^2)$ is some other possible function. Equations 1.13 and 1.14 tell us that x' is zero whenever $x = vt$, and that t' is

---

[6] Once again, this is actually a slight variation of Einstein's approach.

zero whenever $t = vx$. Because of the symmetry in these two equations, the t' axis is just a reflection of the x' axis about the line $x = t$, and vice versa.

What we know so far is that our transformation equations should take the following form:

$$x' = (x - vt)f(v^2)$$

$$t' = (t - vx)g(v^2). \qquad (1.15)$$

Our next task is to figure out what the functions $f(v^2)$ and $g(v^2)$ actually are. To do that we'll consider the path of a light ray in these two frames, and apply Einstein's principle that the speed of light is the same in both of them. If the speed of light c equals 1 in the stationary frame, it must also equal 1 in the moving frame. To rephrase this: If we start out with a light ray that satisfies $x = t$ in the stationary frame, it must also satisfy $x' = t'$ in the moving frame. To put it another way, if

$$x = t,$$

then it must follow that

$$x' = t'.$$

Let's go back to Eqs. 1.15. Setting $x = t$ and requiring $x' = t'$ gives the simple requirement that

$$f(v^2) = g(v^2).$$

In other words the requirement that the speed of light is the same in your frame and my frame leads to the simple condition that the two functions $f(v^2)$ and $g(v^2)$ are the same. Thus we may simplify Eqs. 1.15

$$x' = (x - vt)f(v^2)$$

$$t' = (t - vx)f(v^2). \tag{1.16}$$

To find $f(v^2)$, Einstein used one more ingredient. In effect, he said "Wait a minute, who's to say which frame is moving? Who's to say if my frame is moving relative to you with velocity v, or your frame is moving relative to me with velocity $-v$?" Whatever the relationship is between the two frames of reference, it must be symmetrical. Following this approach, we could invert our entire argument; instead of starting with x and t, and deriving x' and t', we could do exactly the opposite. The only difference would be that for me, you're moving with velocity $-v$, but for you, I'm moving with velocity $+v$. Based on Eqs. 1.16, which express x' and t' in terms of x and t, we can immediately write down the inverse transformations. The equations for x and t in terms of x' and t' are

$$x = (x' + vt')f(v^2)$$

$$t = (t' + vx')f(v^2). \tag{1.17}$$

We should be clear that we wrote Eqs. 1.17 by reasoning about the physical relationship between the two reference frames. They're not just a clever way to solve Eqs. 1.16 for the unprimed variables without doing the work. In fact, now that we have these two sets of equations, we need to verify that they're compatible with each other.

Let's start with Eqs. 1.17 and plug in the expressions for x' and t' from Eqs. 1.16. This may seem circular, but you'll see that it isn't. After making our substitutions, we'll require the results to be equivalent to $x = x$ and $t = t$. How could they be anything else if the equations are valid to begin with? From there, we'll find out the form of $f(v^2)$. The algebra is a little tedious but straightforward. Starting with the first of Eqs. 1.17, the first few substitutions for x unfold as follows:

$$x = (x' + vt')f(v^2)$$

$$x = \{(x - vt)f(v^2) + v(t - vx)f(v^2)\}f(v^2)$$

$$x = (x - vt)f^2(v^2) + v(t - vx)f^2(v^2).$$

Expanding the last line gives

$$x = xf^2(v^2) - v^2xf^2(v^2) - vtf^2(v^2) + vtf^2(v^2),$$

which simplifies to

$$x = xf^2(v^2)(1 - v^2).$$

Canceling x on both sides and solving for $f(v^2)$ gives

$$f(v^2) = \frac{1}{\sqrt{1 - v^2}}. \tag{1.18}$$

Now we have everything we need to transform coordinates in the rest frame to coordinates in the moving frame and vice versa. Plugging into Eqs. 1.16 gives

$$x' = \frac{x - vt}{\sqrt{1 - v^2}} \tag{1.19}$$

$$t' = \frac{t - vx}{\sqrt{1 - v^2}}. \tag{1.20}$$

These are, of course, the famous Lorentz transformations between the rest and moving frames.

Art: *"Wow, that's incredibly clever, Lenny. Did you figure it all out yourself?"*

Lenny: *"I wish. No, I'm simply following Einstein's paper. I haven't read it for fifty years, but it left an impression."*

Art: *"Okay, but how come they're called Lorentz transformations if they were discovered by Einstein?"*

1.2.3 Historical Aside

To answer Art's question, Einstein was not the first to discover the Lorentz transformation. That honor belongs to the Dutch physicist Hendrik Lorentz. Lorentz, and others even before him—notably George FitzGerald—had speculated that Maxwell's theory of electromagnetism required moving objects to contract along the direction of motion, a phenomenon that we now call *Lorentz contraction*. By 1900 Lorentz had written down the Lorentz transformations motivated by this contraction of moving bodies. But the views of Einstein's predecessors were different and in a sense a throwback to older ideas rather than a new starting point. Lorentz and FitzGerald imagined that the interaction between the stationary ether and the moving atoms of all ordinary matter would cause a pressure that would squeeze matter along the direction of motion. To some approximation the pressure would contract all matter by the same amount so that the effect could be represented by a coordinate transformation.

Just before Einstein's paper, the great French mathematician Henri Poincaré published a paper in which he derived the Lorentz transformation from the requirement that Maxwell's equations take the same form in every inertial frame. But none of these works had the clarity, simplicity, and generality of Einstein's reasoning.

1.2.4   Back to the Equations

If we know the coordinates of an event in the rest frame, Eqs. 1.19 and 1.20 tell us the coordinates of the same event in the moving frame. Can we go the other way? In other words, can we predict the coordinates in the rest frame if we know them in the moving frame? To do so we might solve the equations for x and t in terms of x' and t', but there is an easier way.

All we need is to realize that there is a symmetry between the rest and moving frames. Who, after all, is to say which frame is moving and which is at rest? To interchange their roles, we might just interchange the primed and unprimed coordinates in Eqs. 1.19 and 1.20. That's almost correct but not quite.

Consider this: If I am moving to the right relative to you, then you are moving to the left relative to me. That means your velocity relative to me is $-v$. Therefore when I write the Lorentz transformations for x, t in terms of x', t', I will need to replace v with $-v$. The result is

$$x = \frac{x' + vt'}{\sqrt{1 - v^2}} \tag{1.21}$$

$$t = \frac{t' + vx'}{\sqrt{1 - v^2}}. \tag{1.22}$$

Switching to Conventional Units

What if the speed of light is not chosen to be 1? The easiest way to switch from relativistic units back to conventional units is to make sure our equations are dimensionally consistent in those units. For example, the expression $x - vt$ is dimensionally consistent as it is because both x and vt have units of length—say meters. On the other hand $t - vx$ is not dimensionally consistent in conventional units; t has units of seconds and vx has units of meters squared over seconds. There is a unique way to fix the units. Instead of $t - vx$ we replace it with

$$t - \frac{v}{c^2}x.$$

Now both terms have units of time, but if we happen to use units with $c = 1$ it reduces to the original expression $t - vx$.

Similarly the factor in the denominators, $\sqrt{1 - v^2}$, is not dimensionally consistent. To fix the units we replace v with v/c. With these replacements the Lorentz transformations can be written in conventional units:

$$x' = \frac{x - vt}{\sqrt{1 - \frac{v^2}{c^2}}} \tag{1.23}$$

$$t' = \frac{t - \frac{vx}{c^2}}{\sqrt{1 - \frac{v^2}{c^2}}}. \tag{1.24}$$

Notice that when v is very small compared to the speed of light, v^2/c^2 is even smaller. For example, If v/c is $1/10$, then v^2/c^2 is $1/100$. If $v/c = 10^{-5}$, then v^2/c^2 is truly a small number, and the expression $\sqrt{1 - v^2/c^2}$ in the denominator is very close to 1.[7] To a very good approximation we can write

$$x' = x - vt.$$

That's the good old Newtonian version of things. What happens to the time equation (Eq. 1.24) when v/c is very small? Suppose v is 100 meters per second. We know that c is

---

[7] For $v/c = 10^{-5}$ one finds $\sqrt{1 - v^2/c^2} = 0.99999999995$.

very big, about 3×10^8 meters per second. So v/c^2 is a very tiny number. If the velocity of the moving frame is small, the second term in the numerator, vx/c^2, is negligible and to a high degree of approximation the second equation of the Lorentz transformation becomes the same as the Newtonian transformation

$$t' = t.$$

For frames moving slowly relative to each other, the Lorentz transformation boils down to the Newtonian formula. That's a good thing; as long as we move slowly compared to c, we get the old answer. But when the velocity becomes close to the speed of light the corrections become large—huge when v approaches c.

The Other Two Axes

Eqs. 1.23 and 1.24 are the Lorentz transformation equations in common, or conventional, units. Of course, the full set of equations must also tell us how to transform the other two components of space, y and z. We've been very specific about what happens to the x and t coordinates when frames are in relative motion along the x axis. What happens to the y coordinate?

We'll answer this with a simple thought experiment. Suppose your arm is the same length as my arm when we're both at rest in your frame. Then I start moving at constant velocity in the x direction. As we move past each other, we each hold out an arm at a right angle to the direction of our relative motion. Question: As we move past each other, would our arms still be equal in length, or would yours be longer than mine? By the symmetry of this situation, it's clear that our arms are going to match, because there's no reason for one to be longer than the other. Therefore, the rest of the Lorentz transformation must be $y' = y$ and $z' = z$. In other words, interesting things happen only in the x, t plane when the relative motion is along the x axis. The x and t coordinates get mixed up with each other, but y and z are passive.

For easy reference, here's the complete Lorentz transformation in conventional units for a reference frame (the primed frame) moving with velocity v in the positive x direction relative to the unprimed frame:

$$x' = \frac{x - vt}{\sqrt{1 - v^2/c^2}} \tag{1.25}$$

$$t' = \frac{t - vx/c^2}{\sqrt{1 - v^2/c^2}} \tag{1.26}$$

$$y' = y \tag{1.27}$$

$$z' = z. \tag{1.28}$$

1.2.5 Nothing Moves Faster than Light

A quick look at Eqs. 1.25 and 1.26 indicates that something strange happens if the relative velocity of two frames is larger than c. In that case $1 - v^2/c^2$ becomes negative and $\sqrt{1 - v^2/c^2}$ becomes imaginary. That's obvious nonsense. Metersticks and clocks can only define real-valued coordinates.

Einstein's resolution of this paradox was an additional postulate: no material system can move with a velocity greater than light. More accurately, no material system can move faster than light relative to any other material system.

In particular no two observers can move, relative to each other, faster than light.

Thus we never have need for the velocity v to be greater than c. Today this principle is a cornerstone of modern physics. It's usually expressed in the form that no signal can travel faster than light. But since signals are composed of material systems, even if no more substantial than a photon, it boils down to the same thing.

1.3   General Lorentz Transformation

These four equations remind us that we have considered only the simplest kind of Lorentz transformation: a transformation where each primed axis is parallel to its unprimed counterpart, and where the relative motion between the two frames is only along the shared direction of the x and x' axes.

Uniform motion is simple, but it's not always that simple. There's nothing to prevent the two sets of space axes from being oriented differently, with each primed axis at some nonzero angle to its unprimed counterpart.[8] It's also easy to visualize the two frames moving with respect to each other not only in the x direction, but along the y and z directions as well. This raises a question: By ignoring these factors, have we missed something essential about the physics of uniform motion? Happily, the answer is *no*.

Suppose you have two frames in relative motion along some oblique direction, not along any of the coordinate axes. It would be easy to make the primed axes line up with the unprimed axes by performing a sequence of rotations. After

---

[8] We're talking about a *fixed* difference in orientation, not a situation where either frame is rotating with a nonzero angular velocity.

doing those rotations, you would again have uniform motion in the x direction. The general Lorentz transformation— where two frames are related to each other by an arbitrary angle in space, and are moving relative to each other in some arbitrary direction—is equivalent to:

1. A rotation of space to align the primed axes with the unprimed axes.
2. A simple Lorentz transformation along the new x axis.
3. A second rotation of space to restore the original orientation of the unprimed axes relative to the primed axes.

As long as you make sure that your theory is invariant with respect to the *simple* Lorentz transformation along, say, the x axis, *and* with respect to rotations, it will be invariant with respect to any Lorentz transformation at all.

As a matter of terminology, transformations involving a relative velocity of one frame moving relative to another are called *boosts*. For example, the Lorentz transformations like Eqs. 1.25 and 1.26 are referred to as boosts along the x axis.

1.4 Length Contraction and Time Dilation

Special relativity, until you get used to it, is counterintuitive—perhaps not as counterintuitive as quantum mechanics, but nevertheless full of paradoxical phenomena. My advice is that when confronted with one of these paradoxes, you should draw a spacetime diagram. Don't ask your physicist friend, don't email me—draw a spacetime diagram.

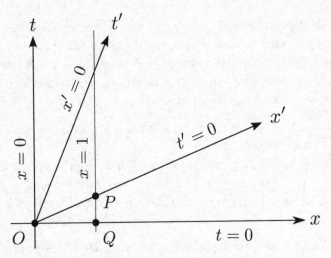

Figure 1.5: Length Contraction.

Length Contraction

Suppose you're holding a meterstick and I'm walking past you in the positive x direction. You know that your stick is 1 meter long, but I'm not so sure. As I walk past, I measure your meterstick relative to the lengths of my metersticks. Since I'm moving, I have to be very careful; otherwise, I could be measuring the end points of your meterstick at two different times. Remember, events that are simultaneous in your frame are not simultaneous in mine. I want to measure the end points of your meterstick at exactly the same time in *my* frame. That's what I *mean* by the length of your meterstick in my frame.

Fig. 1.5 shows a spacetime diagram of this situation. In your frame, the meterstick is represented by a horizontal line segment $\overline{OQ}$ along the x axis, which is a surface of simultaneity for you. The meterstick is at rest, and the world lines of its end points are the vertical lines $x = 0$ and $x = 1$ in your frame.

In my moving frame, that same meterstick at an instant of time is represented by line segment $\overline{OP}$ along the x' axis. The x' axis is a surface of simultaneity for me and is tilted in the diagram. One end of the meterstick is at our common origin O as I pass it. The other end of the stick at time $t' = 0$ is labeled P in the diagram.

To measure the location of both ends at time $t' = 0$ in my frame, I need to know the coordinate values of x' at points O and P. But I already know that x' is zero at point O, so all I need to do is calculate the value of x' at point P. We'll do this the easy way, using relativistic units (speed of light equals 1). In other words, we'll use the Lorentz transformation of Eqs. 1.19 and 1.20.

First, notice that point P sits at the intersection of two lines, $x = 1$ and $t' = 0$. Recall (based on Eq. 1.20) that $t' = 0$ means that $t = vx$. Substituting vx for t in Eq. 1.19 gives

$$x' = \frac{(x - v^2 x)}{\sqrt{1 - v^2}}.$$

Plugging $x = 1$ into the preceding equation gives

$$x' = \frac{(1 - v^2)}{\sqrt{1 - v^2}}$$

or

$$x' = \sqrt{1 - v^2}.$$

So there it is! The moving observer finds that at an instant of time—which means along the surface of simultaneity $t' = 0$—the two ends of the meterstick are separated by distance $\sqrt{1 - v^2}$; in the moving frame, the meterstick is a little *shorter* than it was at rest.

It may seem like a contradiction that the same meterstick has one length in your frame and a different length in my frame. Notice, though, that the two observers are really talking about two different things. For the rest frame, where the meterstick itself is at rest, we're talking about the distance from point O to point Q, as measured by stationary metersticks. In the moving frame, we're talking about the distance between point O and point P, measured by moving measuring rods. P and Q are different points in spacetime, so there's no contradiction in saying that $\overline{OP}$ is shorter than $\overline{OQ}$.

Try doing the opposite calculation as an exercise: Starting with a moving meterstick, find its length in the rest frame. Don't forget—begin by drawing a diagram. If you get stuck you can cheat and continue reading.

Think of a moving meterstick being observed from the rest frame. Fig. 1.6 shows this situation. If the meterstick is 1 unit long in its *own* rest frame, and its leading end passes through point Q, what do we know about its world line? Is it $x = 1$? No! The meterstick is 1 meter long in the moving frame, which means the world line of its leading end is $x' = 1$. The observer at rest now sees the meterstick being the length of line segment $\overline{OQ}$, and the x coordinate of point Q is *not* 1. It's some value calculated by the Lorentz transformation. When you do this calculation, you'll find that this length is also shortened by the factor $\sqrt{1 - v^2}$.

The moving metersticks are short in the stationary frame, and the stationary metersticks are short in the moving frame. There's no contradiction. Once again, the observers are just talking about different things. The stationary observer is talking about lengths measured at an instant of his time. The moving observer is talking about lengths measured at an instant of the *other* time. Therefore, they have different notions of what they mean by length because they have different notions of simultaneity.

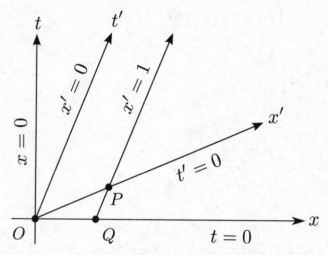

Figure 1.6: Length Contraction Exercise.

Exercise 1.1: Show that the x coordinate of point Q in Fig. 1.6 is $\sqrt{1 - v^2}$.

Time Dilation

Time dilation works pretty much the same way. Suppose I have a moving clock—my clock. Assume my clock is moving along with me at uniform velocity, as in Fig. 1.7.

Here's the question: At the instant when my clock reads $t' = 1$ in my frame, what is the time in your frame? By the way, my standard wristwatch is an excellent timepiece, a Rolex.[9] I want to know the corresponding value of t measured by your Timex. The horizontal surface in the diagram (the dashed line) is the surface that *you* call synchronous. We

---

[9] If you don't believe me, ask the guy who sold it to me on Canal Street for twenty-five bucks.

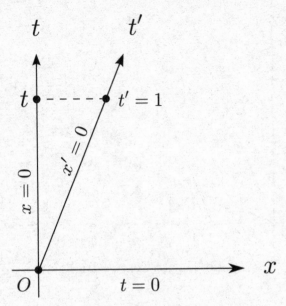

Figure 1.7: Time Dilation.

need two things in order to pin down the value of t in your frame. First, my Rolex moves on the t' axis which is represented by the equation $x' = 0$. We also know that $t' = 1$. To figure out t, all we need is one of the Lorentz transformation equations (Eq. 1.22),

$$t = \frac{t' + vx'}{\sqrt{1 - v^2}}.$$

Plugging in $x' = 0$ and $t' = 1$, we find

$$t = \frac{1}{\sqrt{1 - v^2}}.$$

Because the denominator on the right side is smaller than 1, t itself is bigger than 1. The time interval measured along the t axis (your Timex) is bigger than the time interval measured by the moving observer along the t' axis (my Rolex) by a factor of $1/\sqrt{1-v^2}$. In short, $t > t'$.

To put it another way, as viewed from the rest frame, moving clocks run slower by a factor of $\sqrt{1-v^2}$.

The Twin Paradox

Lenny: *Hey Art! Say hello to Lorentz over here. He has a question.*

Art: Lorentz *has a question for* us?

Lorentz: *Please call me* Lor'ntz. *It's the original Lorentz contraction. In all the years I've been coming to Hermann's Hideaway, I've never seen either one of you guys without the other. Are you biological twins?*

Art: *What? Look, if we were biological twins, either I'd be a genius—don't choke on your sausage, Lor'ntz, it's not that funny. As I was saying, either I'd be a genius or Lenny would be some wiseguy from the Bronx. Wait a minute ...*[10]

Time dilation is the origin of the so-called twin paradox. In Fig. 1.8, Lenny remains at rest, while Art takes a high-speed journey in the positive x direction. At the point labeled $t' = 1$ in the diagram, Art at the age of 1 turns around and heads back home.

We've already calculated the amount of rest frame time that elapses between the origin and the point labeled t in the diagram. It's $1/\sqrt{1-v^2}$. In other words, we find that less time elapses along the path of the moving clock than along the path of the stationary clock. The same thing is true for

---

[10] In what follows we will pretend that Art and Lenny were both born at the same spacetime event (labeled O) in Fig. 1.8.

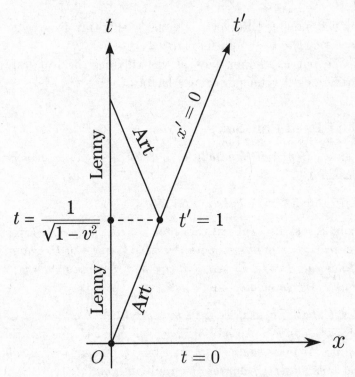

Figure 1.8: Twin Paradox.

the second leg of the journey. When Art returns home, he finds that his twin Lenny is older than himself.

We've calibrated the ages of Art and Lenny by the time registered on their watches. But the same time dilations that slowed Art's watch from the viewpoint of the rest frame would affect any clock, including the biological aging clock. Thus in an extreme case, Art could return home still a boy while Lenny would have a long gray beard.

Two aspects of the twin paradox often leave people confused. First, it seems natural to expect the experiences of the two twins to be symmetrical. If Lenny sees Art moving away from him, then Art also sees Lenny traveling away, but in the opposite direction. There are no preferred directions in space, so why should they age any differently? But in fact, their experiences are not symmetrical at all. The traveling twin undergoes a large acceleration in order to change directions, while the stay-at-home twin does not. This difference is crucial. Because of the abrupt reversal, Art's frame is not a single inertial frame, but Lenny's is. We invite you to develop this idea further in the following exercise.

Exercise 1.2: In Fig. 1.8, the traveling twin not only reverses directions but switches to a different reference frame when the reversal happens.

a) Use the Lorentz transformation to show that before the reversal happens, the relationship between the twins *is* symmetrical. Each twin sees the other as aging more slowly than himself.

b) Use spacetime diagrams to show how the traveler's abrupt switch from one frame to another changes his definition of simultaneity. In the traveler's new frame, his twin is suddenly much older than he was in the traveler's original frame.

Another point of confusion arises from simple geometry. Referring back to Fig. 1.7, recall that we calculated the "time distance" from point O to the point labeled $t' = 1$ to be smaller than the distance from O to the point labeled t ($= 1/\sqrt{1 - v^2}$) along the t axis. Based on these two values, the vertical leg of this right triangle is longer than its hypotenuse. Many people find this puzzling because the numerical comparison seems to contradict the visual message in the diagram. In fact, this puzzle leads us to one of the central ideas in relativity, the concept of an *invariant*. We'll discuss this idea extensively in Section 1.5.

The Stretch Limo and the Bug

Another paradox is sometimes called the Pole in the Barn paradox. But in Poland they prefer to call it the Paradox of the Limo and the Bug.

Art's car is a VW Bug. It's just under 14 feet long. His garage was built to just fit the Bug.

Lenny has a reconditioned stretch limo. It's 28 feet long. Art is going on vacation and renting his house to Lenny, but before he goes the two friends get together to make sure that Lenny's car will fit in Art's garage. Lenny is skeptical, but Art has a plan. Art tells Lenny to back up and get a good distance from the garage. Then step on the gas and accelerate like blazes. If Lenny can get the limo up to 161,080 miles per second before getting to the back end of the garage, it will just fit in. They try it.

Art watches from the sidewalk as Lenny backs up the limo and steps on the gas. The speedometer jumps to 172,000 mps, plenty of speed to spare. But then Lenny looks out at the garage. "Holy cow! The garage is coming at me really fast, and it's less than half its original size! I'll never fit!"

"Sure you will, Lenny. According to my calculation, in the rest frame of the garage you are just a bit longer than thirteen feet. Nothing to worry about."

"Geez, Art, I hope you're right."

Fig. 1.9 is a spacetime diagram including Lenny's stretch limo shown in the dark shaded region, and the garage shown

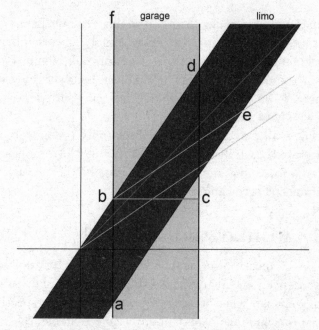

Figure 1.9: Stretch Limo-Garage Spacetime Diagram.

as lightly shaded. The front end of the limo enters the garage at a and leaves (assuming that Art left the back door of the garage open) just above c. The back end of the limo enters at b and leaves at d. Now look at the line $\overline{bc}$. It is part of a surface of simultaneity in the rest frame of the garage, and as you can see, the entire limo is contained in the garage at that time. That's Art's claim: In his frame the limo could be made to fit the garage. But now look at Lenny's surfaces of simultaneity. The line $\overline{be}$ is such a surface, and as you can also see, the limo overflows the garage. As Lenny worried, the limo does not fit.

The figure makes clear what the problem is. To say that the limo is in the garage means that the front and back are simultaneously in the garage. There's that word *simultaneous* again. Simultaneous according to whom, Art? Or Lenny? To say that the car is in the garage simply means different things in different frames. There is no contradiction in saying that at some instant in Art's frame the limo was indeed in the garage—and that at no instant in Lenny's frame was the limo wholly in the garage.

Almost all paradoxes of special relativity become obvious when stated carefully. Watch out for the possibly implicit use of the word *simultaneous*. That's usually the giveaway— simultaneous according to whom?

1.5   Minkowski's World

One of the most powerful tools in the physicist's toolbag is the concept of an invariant. An invariant is a quantity that doesn't change when looked at from different perspectives. Here we mean some aspect of spacetime that has the same value in every reference frame.

To get the idea, we'll take an example from Euclidean geometry. Let's consider a two-dimensional plane with two sets of Cartesian coordinates, x, y and a second set x', y'. Assume that the origins of the two coordinate systems are at the same point, but that the x', y' axes (the primed axes) are rotated counterclockwise by a fixed angle with respect to the unprimed axes. There is no time axis in this example, and there are no moving observers; just the ordinary Euclidean plane of high school geometry. Fig. 1.10 gives you the picture.

Consider an arbitrary point P in this space. The two coordinate systems do not assign the same coordinate values to P. Obviously, the x and the y of this point are not the same numbers as the x' and the y', even though both sets of coordinates refer to the same point P in space. We would say that the coordinates are not invariant.

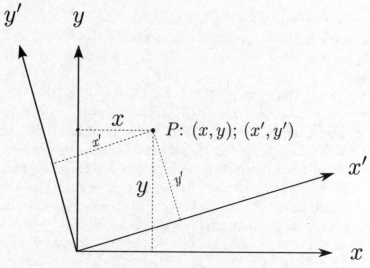

Figure 1.10: Euclidean Plane.

However, there's a property that *is* the same, whether you calculate it in primed or unprimed coordinates: *P*'s *distance from the origin*. That distance is the same in every coordinate system regardless of how it's oriented. The same is true for the square of the distance. To calculate this distance in the unprimed coordinates we use the Pythagorean theorem, $d^2 = x^2 + y^2$, to get the square of the distance. If we use primed coordinates instead, the same distance would be given by $x'^2 + y'^2$. Therefore it follows that

$$x^2 + y^2 = x'^2 + y'^2.$$

In other words, for an arbitrary point *P* the quantity $x^2 + y^2$ is invariant. *Invariant* means that it doesn't depend on which coordinate system you use to work it out. You get the same answer no matter what.

One fact about right triangles in Euclidean geometry is that the hypotenuse is generally larger than either side (unless one side is zero, in which case the hypotenuse is equal to the other side). This tells us that the distance d is at least as large as x or y. By the same argument it is at least as large as x' or y'.

Circling back to relativity, our discussion of the twin paradox involved something that looked a lot like a right triangle. Go back to Fig. 1.8 and consider the triangle formed from the lines connecting the three black dots—the horizontal dashed line, the first half of Lenny's vertical world line, and the hypotenuse formed by the first leg of Art's journey. We can think of the dashed-line distance between the two later dots to define a spacetime distance between them.[11] The time along Lenny's side of the triangle can also be thought of as a spacetime distance. Its length would be $1/\sqrt{1 - v^2}$. And finally the time ticked off during the first half of Art's trip is the spacetime length of the hypotenuse. But a moment's inspection shows something unusual—the vertical leg is longer than the hypotenuse (that's why Lenny had time to grow a beard while Art remained a boy). This immediately tells us that Minkowski space is not governed by the same laws as Euclidean space.

Nevertheless we may ask: Is there an analogous invariant quantity in Minkowski space, associated with the Lorentz transformation—a quantity that stays the same in every inertial reference frame? We know that the square of the distance from the origin to a fixed point P is invariant under simple rotations of Euclidean coordinates. Could a similar quantity, possibly $t^2 + x^2$, be invariant under Lorentz transformations? Let's try it out. Consider an arbitrary point P in a spacetime diagram. This point is characterized by a t value

---

[11] We're using the term *spacetime distance* in a generic sense. Later on, we'll switch to the more precise terms *proper time* and *spacetime interval*.

and an x value, and in some moving reference frame it's also characterized by a t' and an x'. We already know that these two sets of coordinates are related by the Lorentz transformation. Let's see if our guess,

$$t'^2 + x'^2 \stackrel{?}{=} t^2 + x^2,$$

is correct. Using the Lorentz transformation (Eqs. 1.19 and 1.20) to substitute for t' and x', we have

$$t'^2 + x'^2 \stackrel{?}{=} \frac{(t - vx)^2}{1 - v^2} + \frac{(x - vt)^2}{1 - v^2},$$

which simplifies to

$$t'^2 + x'^2 \stackrel{?}{=} \frac{t^2 + v^2 x^2 - 2vtx}{1 - v^2} + \frac{x^2 + v^2 t^2 - 2vtx}{1 - v^2}.$$

Does the right side equal $t^2 + x^2$? No way! You can see immediately that the tx term in the first expression adds to the tx term in the second expression. They do not cancel, and there's no tx term on the left side to balance things out. They can't be the same.

But if you look carefully, you'll notice that if we take the difference of the two terms on the right side rather than their sum, the tx terms would cancel. Let's define a new quantity

$$\tau^2 = t^2 - x^2.$$

The result of subtracting x'^2 from t'^2 gives

$$t'^2 - x'^2 = \frac{t^2 + v^2 x^2 - 2vtx}{1 - v^2} - \frac{x^2 + v^2 t^2 - 2vtx}{1 - v^2}$$

$$= \frac{t^2 + v^2 x^2}{1 - v^2} - \frac{x^2 + v^2 t^2}{1 - v^2}. \tag{1.29}$$

After a bit of rearrangement it is exactly what we want.

$$t'^2 - x'^2 = t^2 - x^2 = \tau^2 \tag{1.30}$$

Bingo! We've discovered an invariant, τ^2, whose value is the same under any Lorentz transformation along the x axis. The square root of this quantity, τ, is called the *proper time*. The reason for this name will become clear shortly.

Up to now we've imagined the world to be a "railroad" in which all motion is along the x axis. Lorentz transformations are all boosts along the x axis. By now you may have forgotten about the other two directions perpendicular to the tracks: directions described by the coordinates y and z. Let's bring them back now. In Section 1.2.4, I explained that the full Lorentz transformation (with $c = 1$) for relative motion along the x axis—a boost along x—has four equations:

$$x' = \frac{x - vt}{\sqrt{1 - v^2}}$$

$$t' = \frac{t - vx}{\sqrt{1 - v^2}}$$

$$y' = y$$

$$z' = z.$$

What about boosts along other axes? As I explained in Section 1.3, these other boosts can be represented as combinations of boosts along x and rotations that rotate the x axis to another direction. As a consequence, a quantity will be invariant under all Lorentz transformations if it is invariant with respect to boosts along x and with respect to rotations of space. What about the quantity $\tau^2 = t^2 - x^2$?

We've seen that it is invariant with respect to x-boosts, but it changes if space is rotated. This is obvious because it involves x but not y and z. Fortunately it is easy to generalize τ to a full-fledged invariant. Consider the generalized version of Eq. 1.30,

$$\tau^2 = t^2 - x^2 - y^2 - z^2. \tag{1.31}$$

Let's first argue that τ is invariant with respect to boosts along the x axis. We've already seen that the term $t^2 - x^2$ is invariant. To that we add the fact that the perpendicular coordinates y and z don't change under a boost along x. If neither $t^2 - x^2$ nor $y^2 + z^2$ change when transforming from one frame to another, then obviously $t^2 - x^2 - y^2 - z^2$ will also be invariant. That takes care of boosts in the x direction.

Now let's see why it does not change if the space axes are rotated. Again the argument comes in two parts. The first is that a rotation of spatial coordinates mixes up x, y, and z but has no effect on time. Therefore t is invariant with respect to rotations of space. Next consider the quantity $x^2 + y^2 + z^2$. A three-dimensional version of Pythagoras's theorem tells us that $x^2 + y^2 + z^2$ is the square of the distance of the point x, y, z to the origin. Again this is something that does not change under a rotation of space. Combining the invariance of time and the invariance of the distance to the origin (under rotations in space), we come to the conclusion that the proper time τ defined by Eq. 1.31 is an invariant that all observers will agree on. This holds not only for observers moving in any direction but observers whose coordinate axes are oriented in any way.

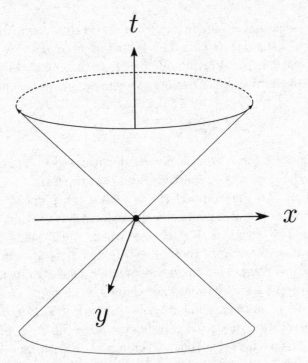

Figure 1.11: Minkowski Light Cone.

1.5.1 Minkowski and the Light Cone

The invariance of the proper time τ is a powerful fact. I don't know if it was known to Einstein, but in the process of writing this section I looked through my ancient worn and faded copy of the Dover edition containing the 1905 paper (the price on the cover was $1.50). I found no mention of Eq. 1.31 or of the idea of spacetime distance. It was Minkowski who first understood that the invariance of proper time, with its unintuitive minus signs, would form the basis for an entirely new four-dimensional geometry of spacetime—Minkowski space. I think it is fair to say that Minkowski deserves the credit for completing in 1908 the special relativity revolution that Einstein set in motion three years earlier. It is to Minkowski that we owe the concept of *time as the fourth dimension* of a four-dimensional spacetime. It still gives me shivers when I read these two papers.

Let's follow Minkowski and consider the path of light rays that start at the origin. Imagine a flash of light—a flashbulb event—being set off at the origin and propagating outward. After a time t it will have traveled a distance ct. We may describe the flash by the equation

$$x^2 + y^2 + z^2 = c^2 t^2. \tag{1.32}$$

The left side of Eq. 1.32 is the distance from the origin, and the right side is the distance traveled by the light signal in time t. Equating these gives the locus of all points reached by the flash. The equation can be visualized, albeit only with three dimensions instead of four, as defining a cone in spacetime. Although he didn't quite draw the cone, he did describe it in detail. Here we draw Minkowski's Light Cone (Fig. 1.11). The upward-pointing branch is called the *future light cone*. The downward-pointing branch is the *past light cone*.

Now let's return to the railroad world of motion strictly along the x axis.

1.5.2 The Physical Meaning of Proper Time

The invariant quantity τ^2 is not just a mathematical abstraction. It has a physical—even an experimental—meaning. To understand it, consider Lenny, as usual, moving along the x axis and Art at rest in the rest frame. They pass each other at the origin O. We also mark a second point D along Lenny's world line, which represents Lenny moving along the t'

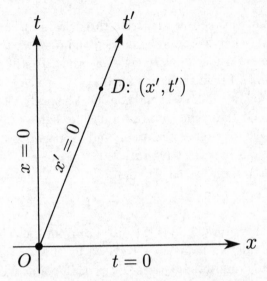

Figure 1.12: Proper Time. Please read the footnote that explains the two meanings of t' in this diagram.

axis.[12] All of this is shown in Fig. 1.12. The starting point along the world line is the common origin O. By definition Lenny is located at $x' = 0$ and he moves along the t' axis.

The coordinates (x, t) refer to Art's frame, and the primed coordinates (x', t') refer to Lenny's frame. The invariant τ^2 is defined as $t'^2 - x'^2$ in Lenny's frame. By definition, Lenny always remains at $x' = 0$ in his own rest frame, and because $x' = 0$ at point D, $t'^2 - x'^2$ is the same as t'^2. Therefore, the equation

$$\tau^2 = t'^2 - x'^2$$

---

[12] We use the label t' in two slightly different ways in this discussion. Its primary meaning is "Lenny's t' coordinate." But we also use it to label the t' axis.

becomes

$$\tau^2 = t'^2,$$

and

$$\tau = t'.$$

But what is t'? It is the amount of time that has passed in Lenny's frame since he left the origin. Thus we find that the invariant τ has a physical meaning:

The invariant proper time along a world line represents the ticking of a clock moving along that world line. In this case it represents the number of ticks of Lenny's Rolex as he moves from the origin to point D.

To complete our discussion of proper time, we write it in conventional coordinates:

$$\tau^2 = t^2 - \frac{x^2}{c^2}.$$

1.5.3 Spacetime Interval

The term *proper time* has a specific physical and quantitative meaning. On the other hand, I've also used the term *space-time distance* as a generic version of the same idea. Going forward, we'll start using a more precise term, *spacetime interval*, $(\Delta s)^2$, defined as

$$(\Delta s)^2 = -\Delta t^2 + (\Delta x^2 + \Delta y^2 + \Delta z^2).$$

To describe the spacetime interval between an event (t, x, y, z) and the origin, we write

$$s^2 = -t^2 + (x^2 + y^2 + z^2).$$

In other words, s^2 is just the negative of τ^2, and is therefore an invariant.[13] So far, the distinction between τ^2 and s^2 has not been important, but it will soon come into play.

1.5.4   Timelike, Spacelike, and Lightlike Separations

Among the many geometric ideas that Minkowski introduced into relativity were the concepts of timelike, spacelike, and lightlike separations between events. This classification may be based on the invariant

$$\tau^2 = t^2 - (x^2 + y^2 + z^2),$$

or on its alter ego

$$s^2 = -t^2 + (x^2 + y^2 + z^2),$$

the spacetime interval that separates an event (t, x, y, z) from the origin. We'll use s^2. The interval s^2 may be negative, positive, or zero, and that's what determines whether an event is timelike, spacelike, or lightlike separated from the origin.

To gain some intuition about these categories, think of a light signal originating at Alpha Centauri at time zero. It takes about four years for that signal to reach us on Earth. In this example, the light flash at Alpha Centauri is at the origin, and we're considering its future light cone (the top half of Fig. 1.11).

---

[13] Sign conventions in relativity are not as consistent as we would like; some authors define s^2 to have the same sign as τ^2.

Timelike Separation

First, consider a point that lies inside the cone. That will be the case if the magnitude of its time coordinate $|t|$ is greater than the spatial distance to the event—in other words, if

$$-t^2 + (x^2 + y^2 + z^2) < 0.$$

Such events are called *timelike* relative to the origin. All points on the t axis are timelike relative to the origin (I'll just call them timelike). The property of being timelike is invariant: If an event is timelike in any frame, it is timelike in all frames.

If an event on Earth happens more than four years after the flash was sent, then it's timelike relative to the flash. Those events will be too late to be struck by the signal. It will have passed already.

Spacelike Separation

Spacelike events are the ones outside the cone.[14] In other words, they're the events such that

$$-t^2 + (x^2 + y^2 + z^2) > 0.$$

For these events the space separation from the origin is larger than the time separation. Again, the spacelike property of an event is invariant.

Spacelike events are too far away for the light signal to reach. Any event on Earth that takes place earlier than four years after the light signal began its journey cannot possibly be affected by the event that created the flash.

---

[14] Once again, we're using the shorthand term "spacelike event" to mean "event that is spacelike separated from the origin."

Lightlike Separation

Finally, there are the events on the light cone; those for which

$$-t^2 + (x^2 + y^2 + z^2) = 0.$$

Those are the points that a light signal, starting at the origin, would reach. A person who is at a lightlike event relative to the origin would see the flash of light.

1.6   Historical Perspective

1.6.1   Einstein

People often wonder whether Einstein's declaration that "c is a law of physics" was based on theoretical insight or prior experimental results—in particular the Michelson-Morley experiment. Of course, we can't be certain of the answer. No one really knows what's in another person's mind. Einstein himself claimed that he was not aware of Michelson's and Morley's result when he wrote his 1905 paper. I think there's every reason to believe him.

Einstein took Maxwell's equations to be a law of physics. He knew that they give rise to wavelike solutions. At age sixteen, he puzzled over what would happen if you moved along with a light ray. The "obvious" answer is that you'd see a static electric and magnetic field with a wavelike structure that doesn't move. Somehow, he knew that was wrong—that it was *not* a solution to Maxwell's equations. Maxwell's equations say that light moves at the speed of light. I'm inclined to believe that, consistent with Einstein's own account, he didn't know of the Michelson-Morley experiment when he wrote his paper.

In modern language we would explain Einstein's reasoning

a little differently. We would say that Maxwell's equations have a *symmetry* of some kind—some set of coordinate transformations under which the equations have the same form in every reference frame. If you take Maxwell's equations, which contain x's and t's, and plug in the old Galilean rules,

$$x' = x - vt$$

$$t' = t,$$

you would find that these equations take a different form in the primed coordinates. They don't have the same form as in the unprimed coordinates.

However, if you plug the Lorentz transformation into Maxwell's equations, the transformed Maxwell equations have exactly the same form in the primed coordinates as in the unprimed coordinates. In modern language, Einstein's great accomplishment was to recognize that the symmetry structure of Maxwell's equations is not the Galileo transformation but the Lorentz transformation. He encapsulated all of this in a single principle. In a sense, he didn't need to actually know Maxwell's equations (though he did know them, of course). All he needed to know is that Maxwell's equations are a law of physics, and that the law of physics requires light to move with a certain velocity. From there he could just work with the motion of light rays.

1.6.2   Lorentz

Lorentz did know about the Michelson-Morley experiment. He came up with the same transformation equations but interpreted them differently. He envisioned them as effects on moving objects caused by their motion through the ether.

Because of various kinds of ether pressures, objects would be squeezed and therefore shortened.

Was he wrong? I suppose you could say that in some way he wasn't wrong. But he certainly didn't have Einstein's vision of a *symmetry* structure—the symmetry required of space and time in order that it agree with the principle of relativity and the motion of the speed of light. Nobody would have said that Lorentz did what Einstein did.[15] Furthermore, Lorentz didn't think it was exact. He regarded the transformation equations as a first approximation. An object moving through a fluid of some kind would be shortened, and the first approximation would be the Lorentz contraction. Lorentz fully expected that the Michelson-Morley experiment was not exact. He thought there would be corrections to higher powers of v/c, and that experimental techniques would eventually become precise enough to detect differences in the velocity of light. It was Einstein who said this is really a law of physics, a principle.

---

[15] Including Lorentz himself, I believe.

Lecture 2

Velocities and 4-Vectors

Art: *That stuff is incredibly fascinating! I feel completely transformed.*

Lenny: *Lorentz transformed?*

Art: *Yeah, positively boosted.*

It's true, when things move at relativistic velocities they become flat, at least from the viewpoint of the rest frame. In fact, as they approach the speed of light they shrink along the direction of motion to infinitely thin wafers, although to themselves they look and feel fine. Can they shrink even past the vanishing point by moving faster than light? Well, no, for the simple reason that no physical object can move faster than light. But that raises a paradox:

Consider Art at rest in the railway station. The train containing Lenny whizzes past him at 90 percent of the speed of light. Their relative velocity is $0.9c$. In the same rail car with Lenny, Maggie is riding her bicycle along the aisle with a velocity of $0.9c$ relative to Lenny. Isn't it obvious that she is moving faster than light relative to Art? In Newtonian physics we would add Maggie's velocity to Lenny's in order to compute her velocity relative to Art. We would find her moving past Art with velocity $1.8c$, almost twice the speed of light. Clearly there is something wrong here.

2.1 Adding Velocities

To understand what's wrong, we will have to make a careful analysis of how Lorentz transformations combine. Our setup now consists of three observers: Art at rest, Lenny moving relative to Art with velocity v, and Maggie moving relative to Lenny with velocity u. We will work in relativistic units with $c = 1$ and assume that v and u are both positive and less than 1. Our goal is to determine how fast Maggie is moving with respect to Art. Fig. 2.1 shows this setup.

There are three frames of reference and three sets of coordinates. Let (x, t) be Art's coordinates in the rest frame of the rail station. Let (x', t') be coordinates in Lenny's frame—the frame at rest in the train. And finally let (x'', t'') be Maggie's coordinates that move with her bicycle. Each

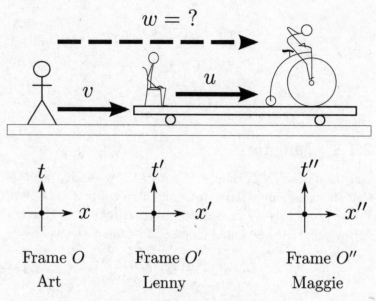

Figure 2.1: Combining Velocities.

pair of frames is related by a Lorentz transformation with the appropriate velocity. For example Lenny's and Art's coordinates are related by

$$x' = \frac{x - vt}{\sqrt{1 - v^2}} \tag{2.1}$$

$$t' = \frac{t - vx}{\sqrt{1 - v^2}}. \tag{2.2}$$

We also know how to invert these relations. This amounts to solving for x and t in terms of x' and t'. I'll remind you that the result is

$$x = \frac{x' + vt'}{\sqrt{1 - v^2}} \tag{2.3}$$

$$t = \frac{t' + vx'}{\sqrt{1 - v^2}}. \tag{2.4}$$

2.1.1   Maggie

Our third observer is Maggie. What we know about Maggie is that she is moving relative to Lenny with relative velocity u. We express this by a Lorentz transformation connecting Lenny's and Maggie's coordinates, this time with velocity u:

$$x'' = \frac{x' - ut'}{\sqrt{1 - u^2}} \tag{2.5}$$

$$t'' = \frac{t' - ux'}{\sqrt{1 - u^2}}. \tag{2.6}$$

Our goal is to find the transformation relating Art's and Maggie's coordinates, and from that transformation read off their relative velocities. In other words, we want to eliminate Lenny.[1] Let's start with Eq. 2.5,

$$x'' = \frac{x' - ut'}{\sqrt{1 - u^2}}.$$

Now, substitute for x' and t' on the right side, using Eqs. 2.1 and 2.2,

$$x'' = \frac{\dfrac{x - vt}{\sqrt{1 - v^2}} - \dfrac{u(t - vx)}{\sqrt{1 - v^2}}}{\sqrt{1 - u^2}},$$

---

[1] Don't panic. It's only Lenny's *velocity* we're trying to get rid of.

and combine the denominators,

$$x'' = \frac{x - vt - u(t - vx)}{\sqrt{1 - v^2}\sqrt{1 - u^2}}. \tag{2.7}$$

Now we come to the main point: determining Maggie's velocity in Art's frame relative to Art. It's not entirely obvious that Eq. 2.7 has the form of a Lorentz transformation (it does), but fortunately we can avoid the issue for the moment. Observe that Maggie's world line is given by the equation $x'' = 0$. For that to be true, we only need to set the numerator of Eq. 2.7 to zero. So let's combine the terms in the numerator,

$$(1 + uv)x - (v + u)t = 0,$$

which results in

$$x = \frac{u + v}{1 + uv} t. \tag{2.8}$$

Now it should be obvious: Eq. 2.8 is the equation for a world line moving with velocity $(u + v)/(1 + uv)$. Thus, calling Maggie's velocity in Art's frame w, we find

$$w = \frac{u + v}{1 + uv}. \tag{2.9}$$

It's now fairly easy to check that Art's frame and Maggie's frame really are related by a Lorentz transformation. I will leave it as an exercise to show that

$$x'' = \frac{x - wt}{\sqrt{1 - w^2}} \tag{2.10}$$

and

$$t'' = \frac{t - wx}{\sqrt{1 - w^2}}. \tag{2.11}$$

To summarize: If Lenny moves with velocity v relative to Art, and Maggie moves with velocity u with respect to Lenny, then Maggie moves with velocity

$$w = \frac{u + v}{1 + uv} \qquad (2.12)$$

with respect to Art. We will analyze this shortly, but first let's express Eq. 2.12 in conventional units by enforcing dimensional consistency. The numerator $u + v$ is dimensionally correct. However, the expression $1 + uv$ in the denominator is not because 1 is dimensionless and both u and v are velocities. We can easily restore dimensions by replacing u and v with u/c and v/c. This gives the relativistic law for the addition of velocities,

$$w = \frac{u + v}{1 + \dfrac{uv}{c^2}}. \qquad (2.13)$$

Let's compare the result with Newtonian expectations. Newton would have said that to find Maggie's velocity relative to Art we should just add u to v. That is indeed what we do in the numerator in Eq. 2.13. But relativity requires a correction in the form of the denominator, $(1 + uv/c^2)$.

Let's look at some numerical examples. First we'll consider the case where u and v are small compared to the speed of light. For simplicity, we'll use Eq. 2.9, where velocities are dimensionless. Just remember that u and v are velocities measured in units of the speed of light. Suppose $u = 0.01$, 1 percent of the speed of light, and v is also 0.01. Plugging these values into Eq. 2.9 gives

$$w = \frac{0.01 + 0.01}{1 + (0.01)(0.01)}$$

or

$$w = \frac{0.02}{1.0001} = 0.019998.$$

The Newtonian answer would, of course, have been 0.02, but the relativistic answer is slightly less. In general, the smaller u and v, the closer will be the relativistic and Newtonian result.

But now let's return to the original paradox: If Lenny's train moves with velocity $v = 0.9$ relative to Art, and Maggie's bicycle moves with $u = 0.9$ relative to Lenny, should we not expect that Maggie is moving faster than light relative to Art? Substituting appropriate values for v and u, we get

$$w = \frac{0.9 + 0.9}{1 + 0.9 \times 0.9}$$

or

$$w = \frac{1.8}{1.81}.$$

The denominator is slightly bigger than 1.8, and the resulting velocity is slightly less than 1. In other words, we have not succeeded in making Maggie go faster than the speed of light in Art's frame.

While we're at it, let's satisfy our curiosity about what would happen if both u and v *equal* the speed of light. We simply find that w becomes

$$w = \frac{1 + 1}{1 + (1)(1)}$$

or

$$w = \frac{2}{2} = 1.$$

Even if Lenny could somehow move at the speed of light relative to Art, and Maggie could move at the speed of light relative to Lenny, still she would still not move faster than light relative to Art.

2.2   Light Cones and 4-Vectors

As we saw in Section 1.5, the proper time

$$\tau^2 = t^2 - (x^2 + y^2 + z^2)$$

and its alter-ego, the spacetime interval relative to the origin,

$$s^2 = -t^2 + (x^2 + y^2 + z^2),$$

are invariant quantities under general Lorentz transformations in four-dimensional spacetime. In other words, these quantities are invariant under any combination of Lorentz boosts and coordinate rotations.[2] We'll sometimes write τ in abbreviated form as

$$\tau^2 = t^2 - \vec{x}^2. \tag{2.14}$$

This is probably the most central fact about relativity.

2.2.1   How Light Rays Move

Back in Lecture 1, we discussed spacetime regions and the trajectories of light rays. Fig. 2.2 illustrates this idea in slightly greater detail. The different kinds of separation correspond to a negative, positive, or zero value of the invariant quantity s. We also discovered the interesting result that if two points in spacetime have a separation of zero, this does *not* mean they have to be the same point. Zero

---

[2] We only showed this explicitly for τ, but the same arguments apply to s.

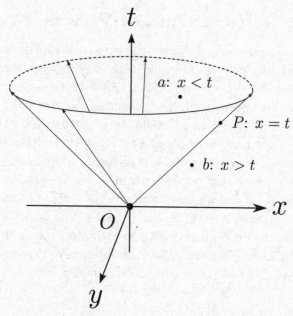

Figure 2.2: Future Light Cone. Relative to the origin: Point a is timelike separated, point b is spacelike separated, and point P is lightlike separated. Only two space dimensions are shown.

separation simply means that the two points are related by the possibility of a light ray going from one of them to the other. That's one important concept of how a light ray moves—it moves in such a way that the proper time (alternately, spacetime interval) along its trajectory is zero. The trajectory of a light ray that starts at the origin serves as a kind of boundary between regions of spacetime that are timelike separated from the origin and those that are spacelike separated.

2.2.2   Introduction to 4-Vectors

We have a reason for bringing the y and z spatial dimensions
back into the picture. The mathematical language of relativity relies heavily on something called a 4-vector, and 4-
vectors incorporate all three dimensions of space. We'll
introduce them here and develop them further in Lecture 3.

The most basic example of a vector in three dimensions is
the interval between two points in space.[3] Given two points,
there's a vector that connects them. It has a direction and a
magnitude. It doesn't matter where the vector begins. If we
move it around, it's still the same vector. You can think of it
as an excursion beginning at the origin and ending at some
point in space. Our vector has coordinates, in this case x, y,
and z, that define the location of the final point.

New Notation

Of course, the names of our coordinates don't necessarily
have to be x, y, and z. We're free to rename them. For
instance, we could call them X^i, with i being 1, 2, or 3. In this
notation, we could write

$$X^i \Longrightarrow (x, y, z).$$

Or we could write

$$X^i \Longrightarrow (X^1, X^2, X^3)$$

instead, and we plan to use that notation extensively. Since
we'll be measuring space *and* time relative to some origin, we
need to add a time coordinate t. As a result, the vector
becomes a four-dimensional object, a 4-vector with one time

---

[3] We're just talking about vectors in space, not the abstract
state vectors of quantum mechanics.

component and three space components. By convention, the time component is first on the list:

$$X^\mu \Longrightarrow (X^0, X^1, X^2, X^3),$$

where (X^0, X^1, X^2, X^3) has the same meaning as (t, x, y, z). Remember, these superscripts are *not* exponents. The coordinate X^3 means "the third space coordinate, the one that's often written as z." It does *not* mean $X \times X \times X$. The distinction between exponents and superscripts should be clear from the context. From now on, when we write four-dimensional coordinates, the time coordinate will come first.

Notice that we're using two slight variants of our index notation:

- X^μ: A Greek index such as μ means that the index ranges over all four values 0, 1, 2, and 3.
- X^i: A Roman index such as i means that the index only includes the three spatial components 1, 2, and 3.

What about the proper time and spacetime interval from the origin? We can write them as

$$\tau^2 = \left(X^0\right)^2 - \left(X^1\right)^2 - \left(X^2\right)^2 - \left(X^3\right)^2$$

and

$$s^2 = -\left(X^0\right)^2 + \left(X^1\right)^2 + \left(X^2\right)^2 + \left(X^3\right)^2.$$

There's no new content here, just notation.[4] But notation is important. In this case, it provides a way to organize our 4-vectors and keep our formulas simple. When you see a μ index, the index runs over the four possibilities of space and time. When you see an i index, the index only runs over

---

[4] You may wonder why we're using superscripts instead of subscripts. Later on (Sec. 4.4.2) we'll introduce subscript notation whose meaning is slightly different.

space. Just as X^i can be thought of as a basic version of a vector in space, X^μ, with four components, represents a 4-vector in spacetime. Just as vectors transform when you rotate coordinates, 4-vectors Lorentz-transform when you go from one moving frame to another. Here is the Lorentz transformation in our new notation:

$$(X')^0 = \frac{X^0 - vX^1}{\sqrt{1 - v^2}}$$

$$(X')^1 = \frac{X^1 - vX^0}{\sqrt{1 - v^2}}$$

$$(X')^2 = X^2$$

$$(X')^3 = X^3.$$

We can generalize this to a rule for the transformation properties of any 4-vector. By definition a 4-vector is any set of components A^μ that transform according to

$$(A')^0 = \frac{A^0 - vA^1}{\sqrt{1 - v^2}}$$

$$(A')^1 = \frac{A^1 - vA^0}{\sqrt{1 - v^2}}$$

$$(A')^2 = A^2$$

$$(A')^3 = A^3 \qquad (2.15)$$

under a boost along the x axis. We also assume that the

spatial components A^1, A^2, A^3 transform as a conventional 3-vector under rotations of space and that A^0 is unchanged.

Just like 3-vectors, 4-vectors can be multiplied by an ordinary number by multiplying the components by that number. We may also add 4-vectors by adding the individual components. The result of such operations is a 4-vector.

4-Velocity

Let's look at another 4-vector. Instead of talking about components relative to an origin, this time we'll consider a small interval along a spacetime trajectory. Eventually, we'll shrink this interval down to an infinitesimal displacement; for now think of it as being small but finite. Fig. 2.3 shows the picture we have in mind. The interval that separates points a and b along the trajectory is ΔX^μ. This simply means the changes in the four coordinates going from one end of a vector to the other. It consists of Δt, Δx, Δy, and Δz.

Now we're ready to introduce the notion of 4-velocity. Four-dimensional velocity is a little different from the normal notion of velocity. Let's take the curve in Fig. 2.3 to be the trajectory of a particle. I'm interested in a notion of velocity at a particular instant along segment $\overline{ab}$. If we were working with an ordinary velocity, we would take Δx and divide it by Δt. Then we would take the limit as Δt approaches zero. The ordinary velocity has three components: the x component, the y component, and the z component. There is no fourth component.

We'll construct the four-dimensional velocity in a similar way. Start with the ΔX^μ. But instead of dividing them by Δt, the ordinary coordinate time, we'll divide them by the proper time $\Delta\tau$. The reason is that $\Delta\tau$ is invariant. Dividing a 4-vector ΔX^μ by an invariant preserves the transformation properties of the 4-vector. In other words, $\Delta X^\mu/\Delta\tau$ is a 4-vector, while $\Delta X^\mu/\Delta t$ is not.

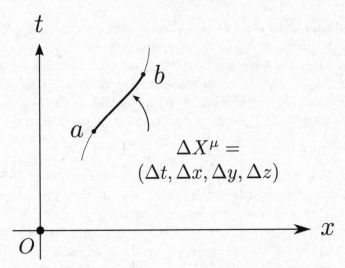

Figure 2.3: Spacetime Trajectory (particle).

In order to distinguish the 4-velocity from the ordinary 3-velocity we'll write it as U instead of V. U has four components U^μ, defined by

$$U^0 = \frac{dX^0}{d\tau} = \frac{dt}{d\tau}$$

$$U^1 = \frac{dX^1}{d\tau} = \frac{dx}{d\tau}$$

$$U^2 = \frac{dX^2}{d\tau} = \frac{dy}{d\tau}$$

$$U^3 = \frac{dX^3}{d\tau} = \frac{dz}{d\tau}. \tag{2.16}$$

We'll have a closer look at 4-velocity in the next lecture. It plays an important role in the theory of the motion of particles. For a relativistic theory of particle motion, we'll need new notions of old concepts such as velocity, position, momentum, energy, kinetic energy, and so on. When we construct relativistic generalizations of Newton's concepts, we'll do it in terms of 4-vectors.

Lecture 3

Relativistic Laws of Motion

Lenny was sitting on a bar stool holding his head in his hands while his cell phone was opened to an email message.

Art: *Whatsa matter, Lenny? Too much beer milkshake?*

Lenny: *Here, Art, take a look at this email. I get a couple like it every day.*

Email Message:[1]

> *Dear Professor Susskino [sic],*
>
> *Einstein made a bad mistake and I discovered it. I wrote to your friend Hawkins [sic] but he didn't answer.*
>
> *Let me explain Einsteins' [sic] mistake. Force equals mass times acceleration. So I push something with a constant force for a long time the acceleration is constant so if I do it long enough the velocity keeps increasing. I calculated that if I push a 220 pound (that's my weight. I*

---

[1] This is an actual email message received 1/22/2007.

should probably go on a diet) person with a continuous
force of 224.809 pounds in a horizontal direction, after a
year he will be moving faster than the speed of light. All I
used was Newtons' [sic] equation F=MA. So Einstein
was wrong since he said that nothing can move faster
than light. I am hoping you will help me publish this as I
am certain that the phycicist's [sic] need to know it. I
have a lot of money and I can pay you.

Art: *Geez, that's awfully stupid.*

By the way, what's wrong with it?

The answer to Art's question is that we're not doing Newton's theory; we're doing Einstein's theory. Physics, including the laws of motion, force, and acceleration, all had to be rebuilt from the ground up, in accord with the principles of special relativity.

We're now ready to tackle that project. We will be especially interested in particle mechanics—how particles move according to special relativity. To accomplish this, we'll need to corral a wide range of concepts, including many from classical mechanics. Our plan is to discuss each idea separately before knitting them all together at the end.

Relativity builds on the classical notions of energy, momentum, canonical momenta, Hamiltonians, and Lagrangians; the principle of least action plays a central role. Though we offer some brief reminders of these ideas as we go, we assume you remember them from the first book of this series, *The Theoretical Minimum: What You Need to Know to Start Doing Physics*. If not, this would be an excellent time to review that material.

3.1   More About Intervals

We discussed timelike and spacelike intervals in Lectures 1 and 2. As we explained, the interval or separation between two points in spacetime is timelike when the invariant quantity

$$(\Delta s)^2 = -(\Delta t)^2 + (\Delta \vec{x})^2 \qquad (3.1)$$

is less than zero, that is, when the time component of the interval is greater than the space component.[2] On the other

---

[2] Remember that when we talk about four-dimensional spacetime—three space coordinates and one time coordinate—the symbol $\Delta \vec{x}$ is a stand-in for all three directions in space. In this context, $(\Delta \vec{x})^2$ refers to the sum of their squares, which we would normally write as $(\Delta x)^2 + (\Delta y)^2 + (\Delta z)^2$.

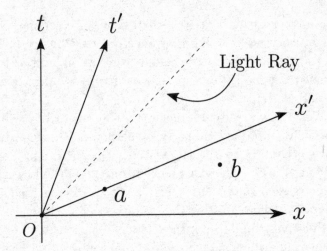

Figure 3.1: Spacelike Interval.

hand, when the spacetime interval $(\Delta s)^2$ between two events is greater than zero, the opposite is true and the interval is called spacelike. This idea was previously illustrated in Fig. 2.2.

3.1.1 Spacelike Intervals

When $(\Delta \vec{x})^2$ is greater than $(\Delta t)^2$, the spatial separation between the two events is greater than their time separation and $(\Delta s)^2$ is greater than zero. You can see this in Fig. 3.1, where the separation between events a and b is spacelike. The line connecting those two points makes an angle smaller than 45 degrees to the x axis.

Spacelike intervals have more space in them than time. They also have the property that you *cannot* find a reference frame in which the two events are located at the same position in space. What you can do instead is find a reference

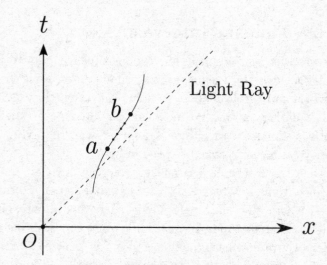

Figure 3.2: Timelike Trajectory.

frame in which they both happen at exactly the same time but at different places. This amounts to finding a frame whose x' axis passes through both points.[3] But there's a much bigger surprise in store. In the t, x frame of Fig. 3.1, event a happens before event b. However, if we Lorentz-transform to the t', x' frame of the diagram, event b happens before event a. Their time order is actually reversed. This brings into sharp focus what we mean by relativity of simultaneity: There's no invariant significance to the idea that one event happens later or earlier than the other if they are spacelike separated.

---

[3] A frame whose x' axis is parallel to that connecting line would serve just as well.

3.1.2   Timelike Intervals

Particles with nonzero mass move along timelike trajectories. To clarify what this means, Fig. 3.2 shows an example. If we follow the path from point a to point b, every little segment of that path is a timelike interval. Saying that a particle follows a timelike trajectory is another way of saying its velocity can never reach the speed of light.

When an interval is timelike, you can always find a reference frame in which the two events happen at the same place—where they have the same space coordinates but occur at different times. In fact, all you need to do is choose a reference frame where the line connecting the two points is at rest—a frame whose t' axis coincides with the line that connects the two points.[4]

3.2   A Slower Look at 4-Velocity

Back in Lecture 2, we introduced some definitions and notation for the 4-velocity. Now it's time to develop that idea further. The components of 4-velocity $dX^\mu/d\tau$ are analogous to the components dX^i/dt of the usual coordinate velocity, except that

- The 4-velocity has—big surprise—four components instead of three, and
- Instead of referring to a rate of change with respect to coordinate time, 4-velocity refers to rate of change with respect to proper time.

As a properly defined 4-vector, the components of 4-velocity

---

[4] A frame whose t' axis is parallel to the connecting line would serve just as well.

transform in the same way as the prototypical 4-vector

$$(t, x, y, z)$$

or

$$(X^0, X^1, X^2, X^3)$$

in our new notation. In other words, their components transform according to the Lorentz transformation. By analogy to ordinary velocities, 4-velocities are associated with small or infinitesimal segments along a path, or a world line, in spacetime. Each little segment has its own associated 4-velocity vector. We'll write the ordinary three-dimensional velocity as

$$\vec{V} = \frac{d\vec{x}}{dt}$$

or

$$V^i = \frac{dX^i}{dt}$$

and the 4-velocity as

$$U^\mu = \frac{dX^\mu}{d\tau}. \tag{3.2}$$

What is the connection between 4-velocity and ordinary velocity? Ordinary non-relativistic velocity has, of course, only three components. This leads us to expect that there's something funny about the fourth (which we label the zeroth) component. Let's start with U^0. Let's write it in the form

$$U^0 = \frac{dX^0}{d\tau} = \frac{dX^0}{dt}\frac{dt}{d\tau}.$$

Now recall that X^0 is just another way of writing t, so that the first factor on the right side is just 1. Thus we can write

$$U^0 = \frac{dt}{d\tau} \tag{3.3}$$

or

$$U^0 = \frac{1}{d\tau/dt}.$$

The next step is to recall that $d\tau = \sqrt{dt^2 - d\vec{x}^2}$ so that

$$\frac{d\tau}{dt} = \frac{\sqrt{dt^2 - d\vec{x}^2}}{dt}$$

or

$$\frac{d\tau}{dt} = \sqrt{1 - \vec{v}^2}, \tag{3.4}$$

where $\vec{v}$ is the usual 3-vector velocity. Now going back to Eq. 3.3 and using Eq. 3.4, we find that

$$\frac{dt}{d\tau} = \frac{1}{\sqrt{1 - \vec{v}^2}} \tag{3.5}$$

and

$$U^0 = \frac{1}{\sqrt{1 - \vec{v}^2}}. \tag{3.6}$$

We see here a new meaning to the ubiquitous factor

$$1/\sqrt{1 - v^2},$$

which appears in Lorentz transformations, Lorentz contraction, and time dilation formulas. We see that it's the time component of the 4-velocity of a moving observer.

What should we make of the time component of U? What would Newton have made of it? Suppose the particle moves very slowly with respect to the velocity of light; in other words, $v << 1$. Then it is clear that U^0 is very close to 1. In the Newtonian limit it is just 1 and carries no special interest. It would have played no role in Newton's thinking. Now let's turn to the spatial components of U. In particular, we can write U^1 as

$$U^1 = \frac{dx}{d\tau},$$

which can also be written as

$$U^1 = \frac{dx}{dt}\frac{dt}{d\tau}.$$

The first factor, dx/dt, is just the ordinary x component of velocity, V^1. The second factor is again given by Eq. 3.5. Putting them together, we have

$$U^i = \frac{V^i}{\sqrt{1 - \vec{v}^2}}.$$

Again, let's ask what Newton would have thought. For very small v, we know that $\sqrt{1 - \vec{v}^2}$ is very close to 1. Therefore the space components of the relativistic velocity are practically the same as the components of the ordinary 3-velocity.

There's one more thing to know about the 4-velocity: Only three of the four components U^μ are independent. They are connected by a single constraint. We can express this in terms of an invariant. Just as the quantity

$$(X^0)^2 - (X^1)^2 - (X^2)^2 - (X^3)^2$$

is an invariant, the corresponding combination of velocity components is invariant, namely

$$(U^0)^2 - (U^1)^2 - (U^2)^2 - (U^3)^2.$$

Is this quantity interesting? I'll leave it as an exercise for you to show that it's always equal to 1,

$$(U^0)^2 - (U^1)^2 - (U^2)^2 - (U^3)^2 = 1. \qquad (3.7)$$

Here's a summary of our results about 4-velocity:

4-Velocity Summary:

$$U^0 = \frac{1}{\sqrt{1 - v^2}} \qquad (3.8)$$

$$U^i = \frac{V^i}{\sqrt{1 - v^2}} \qquad (3.9)$$

$$(U^0)^2 - (\vec{U})^2 = 1 \qquad (3.10)$$

These equations show you how to find the components of 4-velocity. In the nonrelativistic limit where v is close to zero, the expression $\sqrt{1 - v^2}$ is very close to 1, and the two notions of velocity, U^i and V^i, are the same. However, as the speed of an object approaches the speed of light, U^i becomes much bigger than V^i. Everywhere along its trajectory, a particle is characterized by a 4-vector of position X^μ and a 4-vector of velocity U^μ.

Our list of ingredients is nearly complete. We have only one more item to add before we tackle the mechanics of particles.

Exercise 3.1: From the definition of $(\Delta\tau)^2$, verify Eq. 3.7.

3.3 Mathematical Interlude: An Approximation Tool

A physicist's tool kit is not complete without some good approximation methods. The method we describe is an old warhorse that's indispensable despite its simplicity. The basis for the approximation is the binomial theorem.[5] I won't quote the theorem in its general form; just a couple of examples will do. What we will need is a good approximation to expressions like

$$(1 + a)^p$$

that's accurate when a is much smaller than 1. In this expression, p can be any power. Let's consider the example $p = 2$. Its exact expansion is

$$(1 + a)^2 = 1 + 2a + a^2.$$

When a is small, the first term (in this case 1) is fairly close to the exact value. But we want to do a little bit better. The next approximation is

$$(1 + a)^2 \approx 1 + 2a.$$

Let's try this out for $a = .1$ The approximation gives $(1 + .1)^2 \approx 1.2$ whereas the exact answer is 1.21. The smaller we make a, the less important is the a^2 term and the better the approximation. Let's see what happens when $p = 3$. The exact expansion is

$$(1 + a)^3 = 1 + 3a + 3a^2 + a^3$$

---

[5] Look it up if it's not familiar. Here is a good reference: https://en.wikipedia.org/wiki/Binomial_approximation.

and the first approximation would be

$$(1 + a)^3 \approx 1 + 3a.$$

For $a = .1$ the approximation would give 1.3, while the exact answer is 1.4641. Not bad but not great. But now let's try it for $a = .01$. The approximate answer is

$$(1.01)^3 \approx 1.03,$$

while the exact answer is

$$(1.01)^3 = 1.030301,$$

which is much better.

Now, without any justification, I will write the general answer for the first approximation for any value of p:

$$(1 + a)^p \approx 1 + ap. \tag{3.11}$$

In general, if p is not an integer, the exact expression is an infinite series. Nevertheless, Eq. 3.11 is highly accurate for small a and gets better and better as a becomes smaller.

We'll use Eq. 3.11 here to derive approximations for two expressions that show up all the time in relativity theory:

$$\sqrt{1 - v^2} \tag{3.12}$$

and

$$\frac{1}{\sqrt{1 - v^2}}, \tag{3.13}$$

where v represents the velocity of a moving object or reference frame. We do this by writing Eqs. 3.12 and 3.13 in the form

$$\sqrt{1 - v^2} = (1 - v^2)^{1/2}$$

$$\frac{1}{\sqrt{1 - v^2}} = (1 - v^2)^{-1/2}.$$

In the first case the roles of a and p are played by $a = -v^2$ and $p = 1/2$. In the second case $a = -v^2$ but $p = -1/2$. With those identifications our approximations become

$$\sqrt{1 - v^2} \approx 1 - \frac{1}{2} v^2 \tag{3.14}$$

$$\frac{1}{\sqrt{1 - v^2}} \approx 1 + \frac{1}{2} v^2. \tag{3.15}$$

We've written these expressions in relativistic units, so that v is a dimensionless fraction of the speed of light. In conventional units they take the form

$$\sqrt{1 - (v/c)^2} \approx 1 - \frac{1}{2} \frac{v^2}{c^2} \tag{3.16}$$

$$\frac{1}{\sqrt{1 - (v/c)^2}} \approx 1 + \frac{1}{2} \frac{v^2}{c^2}. \tag{3.17}$$

Let me pause for a moment to explain why we're doing this. Why approximate when it's not difficult, especially with modern calculators, to calculate the exact expression to extremely high precision? We're not doing it to make calculations easy (though it may have that effect). We are constructing a new theory for describing motion at very large velocities. However, we are not free to do anything we like; we are constrained by the success of the older theory—Newtonian mechanics—for describing motion much slower than the speed of light. Our real purpose in approximations

like Eqs. 3.16 and 3.17 is to show that the relativistic
equations approach the Newtonian equations when v/c is
very small. For easy reference, here are the approximations
we will use.

Approximations:

$$\sqrt{1 - v^2} \approx 1 - \frac{v^2}{2} \tag{3.18}$$

$$\frac{1}{\sqrt{1 - v^2}} \approx 1 + \frac{v^2}{2} \tag{3.19}$$

From now on we will dispense with the approximation
symbol $\approx$ and use the approximate formulas only when
they are accurate enough to merit an equal sign.

3.4 Particle Mechanics

With all these ingredients in place, we're ready to talk about
particle mechanics. The word *particle* often conjures up the
image of elementary particles such as electrons. However,
we're using the word in a much broader sense. A particle can
be anything that holds itself together. Elementary particles
certainly meet this criterion, but many other things do as
well: the Sun, a doughnut, a golf ball, or my email corres-
pondent. When we speak of the position or velocity of a
particle, what we really mean is the position or velocity of its
center of mass.

Before beginning the next section I urge you to refresh
your knowledge of the principle of least action, Lagrangian
mechanics, and Hamiltonian mechanics if you have forgotten
them. *Volume I* of the Theoretical Minimum series is one
place to find them.

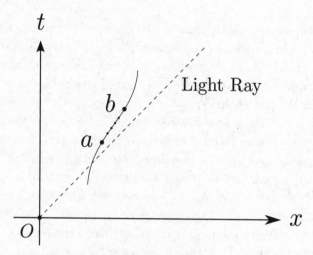

Figure 3.3: Timelike Particle Trajectory.

3.4.1 Principle of Least Action

The principle of least action and its quantum mechanical generalization may be the most central idea in all of physics. All the laws of physics, from Newton's laws of motion, to electrodynamics, to the modern so-called gauge theories of fundamental interactions, are based on the action principle. Do we know exactly why? I think the roots of it are in quantum theory, but suffice it to say that it is deeply connected with energy conservation and momentum conservation. It guarantees the internal mathematical consistency of equations of motion. I discussed action in great detail in the first book of this series. This will be a quick and abbreviated review.

Let's briefly review how the action principle determines the motion of a particle in classical mechanics. The action is a quantity that depends on the trajectory of the particle as it moves through spacetime. You can think of this trajectory as a world line, such as that shown in Fig. 3.2, which we've reproduced here as Fig. 3.3 for convenience. This diagram is a good model for our discussion of action. However, keep in mind that when plotting the position of a system, the x axis represents the entire spatial description of the system—x could stand for a one-dimensional coordinate, but it could also stand for a spatial 3-vector. (It could even represent all the spatial coordinates of a large number of particles, but here we consider only a single particle). We just call it *space* or *coordinate space*. As usual, the vertical axis represents coordinate time, and the trajectory of a system is a curve.

For a particle the Lagrangian depends on the position and velocity of the particle and, most important, it is built out of the kinetic and potential energy. The curve in Fig. 3.3 represents the timelike world line of a single particle with nonzero mass. We'll study its behavior as it goes from point a to point b.[6]

Our development of the least action principle will closely parallel what we did in classical mechanics. The only real difference is that we now add the requirement of frame independence; we want our laws of physics to be the same in every inertial reference frame. We can achieve that by casting our laws in terms of quantities that are the same in every reference frame. In other words, the action should be invariant, and that's best accomplished if its constituents are invariant.

The least action principle says that if a system starts out at point a and ends up at point b, it will "choose" a particular

------

[6] The concept of a world line is just as good in nonrelativistic physics as it is in relativistic physics. But it acquires a certain primacy in relativity because of the connection between space and time—the fact that space and time morph into each other under a Lorentz transformation.

kind of path among all the possible paths. Specifically, it chooses the path that minimizes the quantity we call action.[7] Action is built up incrementally as a sum over the trajectory. Each little segment of the trajectory has a quantity of action associated with it. We calculate the action over the entire path from a to b by adding all these little chunks of action together. As we shrink the segments down to infinitesimal size, the sum becomes an integral. This idea of action being an integral over a trajectory with fixed end points is something we take directly from prerelativity physics. The same is true for the idea that the system somehow chooses the path that minimizes the action integral.

3.4.2 A Quick Review of Nonrelativistic Action

Recall the formula for the action of a nonrelativistic particle from *Volume I*. The action is an integral along the trajectory of a system, and the integrand is called the Lagrangian, denoted $\mathcal{L}$. In symbols,

$$Action = \int_a^b \mathcal{L}\, dt. \tag{3.20}$$

As a rule, the Lagrangian is a function of the position and velocity along the trajectory. In the simplest case of a particle with no forces acting on it, the Lagrangian is just the kinetic

---

[7] There is a technical point that I will mention in order to preempt complaints by sophisticated readers. It is not strictly true that the action needs to be minimum. It may be minimum, maximum, or more generally stationary. Generally this fine point will play no role in the rest of this book. We will therefore follow tradition and refer to the principle of least (minimum) action.

energy $\frac{1}{2}mv^2$. In other words,

$$\mathcal{L} = \frac{1}{2}mv^2$$

$$= \frac{1}{2}m(\dot{x}^2 + \dot{y}^2 + \dot{z}^2), \tag{3.21}$$

where m is the mass of the particle and v is the instantaneous velocity.[8] The action for the nonrelativistic particle is

$$Action = m \int_a^b \frac{1}{2}v^2\,dt. \tag{3.22}$$

Notice that the action is proportional to the mass: For a marble and a bowling ball moving on the same trajectory, the action of the bowling ball is larger than the action of the marble by the ratio of their masses.

3.4.3 Relativistic Action

The nonrelativistic description of particle motion is highly accurate for particles moving much slower than light, but breaks down badly for relativistic particles with larger velocities. To understand relativistic particles we need to start over from the ground up, but one thing stays the same. The theory of relativistic motion is based on the action principle.

How then to calculate the action of a relativistic particle for each little segment along the trajectory? To make sure the laws of motion are the same in every reference frame, the action should be invariant. But there's really only one thing that's invariant when a particle moves from some position to a neighboring position: the proper time separating the two.

---

[8] Recall that a variable with a dot on top of it means "derivative with respect to time." For example, $\dot{x}$ is shorthand for dx/dt.

The proper time from one point to another is a quantity that all observers will agree on. They will not agree on the Δt's or the $\Delta \vec{x}$'s, but they will agree on $\Delta \tau$. So a good guess, and it's the right guess, is to take the action to be *proportional* to the sum of all the little $\Delta \tau$'s. That sum is simply the total proper time along the world line. In mathematical language,

$$Action = -constant \times \sum \Delta \tau$$

where the sum is from one end of a trajectory to the other—from point a to point b in Fig. 3.3. We'll come back to the constant factor and the minus sign shortly.

Once the action has been constructed, we do exactly the same thing as in classical mechanics: Holding the two end points fixed, we wiggle the connecting path around until we find the path that produces the smallest action. Because the action is built up from the invariant quantities $\Delta \tau$, every observer will agree about which path minimizes it.

What is the meaning of the constant factor in the action? To understand it, let's go back to the nonrelativistic case (Eq. 3.22) where we saw that the action for a given path is proportional to the mass of the particle. If we wish to reproduce standard nonrelativistic physics in the limit of small velocities, the relativistic action will also have to be proportional to the mass of the particle. The reason for the minus sign will become clear as we proceed. Let's try defining the action as

$$Action = -m \sum \Delta \tau .$$

Now let's imagine each little segment along the trajectory shrinking down to infinitesimal size. Mathematically, this means we convert our sum to an integral,

$$Action = -m \int_a^b d\tau,$$

and $\Delta\tau$ has become its infinitesimal counterpart $d\tau$. We've added the limits a and b to show that the integral runs from one end of the trajectory to the other. We already know from Eq. 3.4 that

$$\frac{d\tau}{dt} = \sqrt{1 - v^2},$$

and we can use this to replace $d\tau$ in the action integral with $dt\sqrt{1 - v^2}$, resulting in

$$Action = -m \int_a^b dt\sqrt{1 - v^2}.$$

In our new notation, v^2 becomes $(\dot{X}^i)^2$, and the action integral becomes

$$Action = -m \int_a^b dt\sqrt{1 - (\dot{X}^i)^2}. \qquad (3.23)$$

I'm using the symbol $(\dot{X}^i)^2$ to mean $\dot{x}^2 + \dot{y}^2 + \dot{z}^2$, where the dot means derivative with respect to ordinary coordinate time. We might also write that $\dot{x}^2 + \dot{y}^2 + \dot{z}^2 = v^2$ with $\vec{v}$ being the ordinary three-dimensional velocity vector.

We've converted the action integral to something almost familiar: an integral of a function of velocity. It has the same general form as Eq. 3.20, but now instead of the Lagrangian being the nonrelativistic kinetic energy of Eq. 3.21, it has the slightly more complicated form

$$\mathcal{L} = -m\sqrt{1 - (\dot{X}^i)^2} \qquad (3.24)$$

or

$$\mathcal{L} = -m\sqrt{1 - v^2}. \tag{3.25}$$

Before getting more familiar with this Lagrangian, let's put back the appropriate factors of c to restore the correct dimensions for conventional units. To make the expression $1 - (\dot{X}^i)^2$ dimensionally consistent, we must divide the velocity components by c. In addition, to make the Lagrangian have units of energy, we need to multiply the whole thing by c^2. Thus in conventional units,

$$\mathcal{L} = -mc^2\sqrt{1 - \frac{v^2}{c^2}}, \tag{3.26}$$

or in explicit detail,

$$\mathcal{L} = -mc^2\sqrt{1 - \frac{\dot{x}^2 + \dot{y}^2 + \dot{z}^2}{c^2}}. \tag{3.27}$$

In case you didn't notice, Eq. 3.26 marks the first appearance of the expression mc^2.

3.4.4 Nonrelativistic Limit

We would like to show that in the limit of small velocities, the behavior of relativistic systems reduces to Newtonian physics. Since everything about that motion is encoded in the Lagrangian, we only need to show that for small velocities the Lagrangian reduces to Eq. 3.21. It was exactly for this and other similar purposes that we introduced the approximations of Eqs. 3.16 and 3.17,

$$\sqrt{1 - (v/c)^2} \approx 1 - \frac{1}{2}\frac{v^2}{c^2}$$

$$\frac{1}{\sqrt{1 - (v/c)^2}} \approx 1 + \frac{1}{2}\frac{v^2}{c^2}.$$

If we apply the first of these to Eq. 3.26, the result is

$$\mathcal{L} = -mc^2\left(1 - \frac{1}{2}\frac{v^2}{c^2}\right),$$

which we may rewrite as

$$\mathcal{L} = \frac{1}{2}mv^2 - mc^2.$$

The first term, $\frac{1}{2}mv^2$, is good old kinetic energy from New-tonian mechanics: exactly what we expect the nonrelativistic Lagrangian to be. Incidentally, had we not put the overall minus sign in the action, we would not have reproduced this term with the correct sign.

What about the additional term $-mc^2$? Two questions come to mind. The first is whether it makes any difference to the motion of a particle. The answer is that the addition (or subtraction) of a constant to a Lagrangian, for any system, makes no difference at all. We may leave it or delete it without any consequences for the motion of the system. The second question is, what does that term have to do with the equation $E = mc^2$? We'll see that shortly.

3.4.5   Relativistic Momentum

Momentum is an extremely important concept in mechanics, not least of all because it is conserved for a closed system. Moreover, if we divide a system into parts, the rate of change of the momentum of a part is the force on that part due to the rest of the system.

Momentum, often denoted by $\vec{P}$, is a 3-vector that in Newtonian physics is given by the mass times the velocity,

$$\vec{P} = m\vec{v}.$$

Relativistic physics is no different; momentum is still conserved. But the relation between momentum and velocity is more complicated. In his 1905 paper, Einstein worked out the relation in a classic argument that was not only brilliant but characteristically simple. He began by considering an object in the rest frame, and then imagined the object split into two lighter objects, each of which moved so slowly that the whole process could be understood by Newtonian physics. Then he imagined observing that same process from another frame in which the initial object was moving with a large relativistic velocity. The velocity of the final objects could easily be determined by boosting (Lorentz-transforming) the known velocities in the original frame. Then, putting the pieces together, he was able to deduce the expression for their momenta by assuming momentum conservation in the moving frame.

I will use a less elementary, perhaps less beautiful argument, but one that is more modern and much more general. In classical mechanics (go back to *Volume I*) the momentum of a system—say a particle—is the derivative of the Lagrangian with respect to the velocity. In terms of components,

$$P^i = \frac{\partial \mathcal{L}}{\partial \dot{X}^i}. \tag{3.28}$$

For reasons we have not yet explained, we often prefer to

write equations of this kind as

$$P_i = \frac{\partial \mathcal{L}}{\partial \dot{X}^i}, \tag{3.29}$$

where P_i has a subscript instead of a superscript; we've written it here only for reference. We'll explain the meaning of upper and lower indices in Section 5.3.

To find the relativistic expression for momentum of a particle, all we have to do is to apply Eq. 3.28 to the Lagrangian in Eq. 3.27. For example, the x component of momentum is

$$P^x = \frac{\partial \mathcal{L}}{\partial \dot{x}}.$$

Carrying out the derivative, we get

$$P^x = m \frac{\dot{x}}{\sqrt{1 - \dfrac{\dot{x}^2 + \dot{y}^2 + \dot{z}^2}{c^2}}},$$

or more generally,

$$P^i = \frac{mV^i}{\sqrt{1 - \dfrac{v^2}{c^2}}}. \tag{3.30}$$

Let's compare this formula with the nonrelativistic version

$$P^i = mV^i.$$

The first interesting fact is that they are not so different. Go back to the definition of relativistic velocity,

$$U^i = \frac{dX^i}{d\tau}.$$

Comparing Eq. 3.9 with Eq. 3.30 we see that the relativistic momentum is just the mass times the relativistic velocity,

$$P^i = m\frac{dX^i}{d\tau} = mU^i. \qquad (3.31)$$

We probably could have guessed Eq. 3.31 without doing all that work. However, it's important that we can derive it from the basic principles of mechanics: the fundamental definition of momentum as the derivative of the Lagrangian with respect to the velocity.

As you might expect, the relativistic and nonrelativistic definitions of momentum come together when the velocity is much smaller than the speed of light, in which case the expression

$$\frac{1}{\sqrt{1 - \dfrac{v^2}{c^2}}}$$

is very close to 1. But notice what happens as the velocity increases and gets close to c. In that limit the expression blows up. Thus as the velocity approaches c, the momentum of a massive object becomes infinite!

Let's return to the email message that I began this lecture with and see if we can answer the writer. The whole argument rested on Newton's second law: as written by the emailer, $F = MA$. It's well known to readers of the first volume that this can be expressed another way: namely, force is the rate of change of momentum,

$$F = \frac{dP}{dt}. \qquad (3.32)$$

The two ways of expressing Newton's second law are the same within the restricted domain of Newtonian mechanics, where the momentum is given by the usual nonrelativistic formula $P = mV$. However, the more general principles of mechanics imply that Eq. 3.32 is more fundamental and applies to relativistic problems as well as nonrelativistic ones.

What happens if, as the emailer suggested, a constant force is applied to an object? The answer is that the momentum increases uniformly with time. But since getting to the speed of light requires an *infinite* momentum, it will take forever to get there.

3.5 Relativistic Energy

Let's now turn to the meaning of energy in relativistic dynamics. As I'm sure you are aware, it is another conserved quantity. As you probably also know, at least if you have read *Volume I*, energy is the Hamiltonian of the system. If you need a refresher on Hamiltonians, now is the time to take a break and go back to *Volume I*.

The Hamiltonian is a conserved quantity. It's one of the key elements of the systematic approach to mechanics developed by Lagrange, Hamilton, and others. The framework they established allows us to reason from basic principles rather than just make things up as we go along. The Hamiltonian H is defined in terms of the Lagrangian. The most general way to define it is

$$H = \sum_i \dot{Q}^i P^i - \mathcal{L}, \tag{3.33}$$

where the Q^i and P^i are the coordinates and canonical momenta that define the phase space of the system in question. For a moving particle, the coordinates are simply the three components of position, X^1, X^2, X^3, and Eq. 3.33 takes the form

$$H = \sum_i \dot{X}^i P^i - \mathcal{L}. \tag{3.34}$$

We already know from Eq. 3.31 that the momenta are

$$P^i = mU^i,$$

or

$$P^i = \frac{m\dot{X}^i}{\sqrt{1 - v^2}}.$$

We also know from Eq. 3.24 that the Lagrangian is

$$\mathcal{L} = -m\sqrt{1 - (\dot{X}^i)^2}$$

or

$$\mathcal{L} = -m\sqrt{1 - v^2}.$$

Substituting these equations for P^i and $\mathcal{L}$ into Eq. 3.34 results in

$$H = \sum_i \frac{m(\dot{X}^i)^2}{\sqrt{1 - v^2}} + m\sqrt{1 - v^2}.$$

This equation for the Hamiltonian looks like a mess, but we can simplify it quite a bit. First, notice that $(\dot{X}^i)^2$ is just the velocity squared. As a result, the first term does not even need to be written as a sum; it's just $mv^2/\sqrt{1 - v^2}$. If we multiply and divide the second term by $\sqrt{1 - v^2}$, it will have the same denominator as the first term. The resulting numerator is $m(1 - v^2)$. Putting these things together gives us

$$H = \frac{mv^2}{\sqrt{1 - v^2}} + \frac{m(1 - v^2)}{\sqrt{1 - v^2}}.$$

Now it's much simpler, but we're still not done. Notice that mv^2 in the first term cancels mv^2 in the second term, and the whole thing boils down to

$$H = \frac{m}{\sqrt{1 - v^2}}. \qquad (3.35)$$

That's the Hamiltonian—the energy. Do you recognize the factor $1/\sqrt{1 - v^2}$ in this equation? If not, just refer back to Eq. 3.8. It's U^0. Now we can be sure that the zeroth component of the 4-momentum,

$$P^0 = mU^0, \qquad (3.36)$$

is the energy. This is actually a big deal, so let me shout it out loud and clear:

The three components of spatial momentum P^i together with the energy P^0 form a 4-vector.

This has the important implication that the energy and momentum get mixed up under a Lorentz transformation. For example, an object at rest in one frame has energy but no momentum. In another frame the same object has both energy and momentum.

Finally, the prerelativistic notion of momentum conservation becomes the conservation of 4-momentum: the conservation of x-momentum, y-momentum, z-momentum, and energy.

3.5.1   Slow Particles

Before going on, we should figure out how this new concept of energy is related to the old concept. We'll change back to conventional units for a while and put c back into our equations. Take Eq. 3.35. Recognizing that the Hamiltonian is the same thing as the energy, we can write

$$E = \frac{m}{\sqrt{1 - v^2}}. \tag{3.37}$$

To restore the appropriate factors of c we first note that energy has units of mass times velocity squared (an easy way to remember this is from the nonrelativistic expression for kinetic energy $\frac{1}{2}mv^2$). Therefore the right side needs a factor of c^2. In addition the velocity should be replaced with v/c, resulting in

$$E = \frac{mc^2}{\sqrt{1 - v^2/c^2}}. \tag{3.38}$$

Eq. 3.38 is the general formula for the energy of a particle of mass m in terms of its velocity. In that sense it is similar to the nonrelativistic formula for kinetic energy. In fact, we should expect that when the velocity is much less than c, it should reduce to the nonrelativistic formula. We can check that by using the approximation in Eq. 3.19. For small v/c we get

$$E = mc^2 + \frac{mv^2}{2}. \tag{3.39}$$

The second term on the right side of Eq. 3.39 is the nonrelativistic kinetic energy, but what is the first term? It is, of course, a familiar expression, maybe the most familiar in all of physics, namely mc^2. How should we understand its presence in the expression for energy?

Even before the advent of relativity it was understood that the energy of an object is not just kinetic energy. Kinetic energy is the energy due to the motion of an object, but even when the object is at rest it may contain energy. That energy was thought of as the energy needed to assemble the system. What is special about the energy of assembly is that it does not depend on the velocity. We may think of it as "rest energy." Eq. 3.39, which is a consequence of Einstein's theory of relativity, tells us the exact value of the rest energy of any object. It tells us that when the velocity of an object is zero, its energy is

$$E = mc^2. \tag{3.40}$$

I'm sure that this is not the first time any of you have seen this equation, but it may be the first time you've seen it derived from first principles. How general is it? The answer is very general. It does not matter whether the object is an elementary particle, a bar of soap, a star, or a black hole. In a frame in which the object is at rest, its energy is its mass times the square of the speed of light.

Terminology: Mass and Rest Mass

The term *rest mass* is an anachronism, despite its continued use in many undergraduate textbooks. Nobody I know who does physics uses the term *rest mass* anymore. The new convention is that the word *mass* means what the term *rest mass* used to mean.[9] The mass of a particle is a tag that goes with the particle and characterizes the particle itself, not the motion of the particle. If you look up the mass of an electron, you won't get something that depends on whether the electron is moving or stationary. You'll get a number that characterizes the electron at rest. What about the thing that used to be called mass, the thing that does depend on particle motion? We call that *energy*, or perhaps energy divided by the speed of light squared. Energy characterizes a moving particle. Energy at rest is just called mass. We will avoid the term *rest mass* altogether.

---

[9] I believe this "new" convention started about forty to fifty years ago.

3.5.2 Massless Particles

So far we have discussed the properties of massive particles—particles that when brought to rest have a nonzero rest energy. But not all particles have mass. The photon is an example. Massless particles are a little strange. Eq. 3.37 tells us that energy is $m/\sqrt{1-v^2}$. But what is the velocity of a massless particle? It's 1! We're in trouble! On the other hand, perhaps the trouble is not so bad because the numerator and denominator are *both* zero. That doesn't tell us the answer, but at least it leaves some room for negotiation.

Our zero-over-zero conundrum does contain one little seed of wisdom: It's a bad idea to think about the energy of a massless particle in terms of its velocity, because all massless particles move with exactly the *same* velocity. Can they have different energies if they all move with the same velocity? The answer is yes, and the reason is that zero over zero is not determined.

If trying to distinguish massless particles by their velocities is a dead-end road, what can we do instead? We can write their energy as a function of *momentum*.[10] We actually do that quite often in prerelativity mechanics when we write the kinetic energy of a particle. Another way to write

$$E = \frac{1}{2}mv^2$$

is

$$E = \frac{p^2}{2m}.$$

To find the relativistic expression for energy in terms of momentum, there's a simple trick: We use the fact that U^0,

---

[10] In fact, we can think about any particle in this way.

U^x, U^y, and U^z are not completely independent. We worked out their relationship in Section 3.2. Expanding Eq. 3.10 and setting $c = 1$ for the moment, we can write

$$(U^0)^2 - (U^x)^2 - (U^y)^2 - (U^z)^2 = 1.$$

The components of momentum are the same as the components of 4-velocity except for a factor of mass, and we can multiply the preceding equation by m^2 to get

$$m^2(U^0)^2 - m^2(U^x)^2 - m^2(U^y)^2 - m^2(U^z)^2 = m^2. \quad (3.41)$$

We recognize the first term as $(P^0)^2$. But P^0 itself is just the energy, and the remaining three terms on the left side of Eq. 3.41 are the x, y, and z components of the 4-momentum. In other words, we can rewrite Eq. 3.41 as

$$E^2 - P^2 = m^2. \quad (3.42)$$

We can see that Eqs. 3.10 and 3.42 are equivalent to each other. The terms in Eq. 3.10 are components of the 4-velocity, while the terms in Eq. 3.42 are the corresponding components of the 4-momentum. We can solve Eq. 3.42 for E to get

$$E = \sqrt{P^2 + m^2}. \quad (3.43)$$

Now let's put the speed of light back into this equation and see what it looks like in conventional units. I'll leave it as an exercise to verify that the energy equation becomes

$$E = \sqrt{P^2 c^2 + m^2 c^4}. \quad (3.44)$$

Here is our result. Eq. 3.44 gives energy in terms of momentum and mass. It describes all particles whether their masses are zero or nonzero. From this formula we can *immediately* see the limit as the mass goes to zero. We had trouble describing the energy of a photon in terms of its velocity, but we have no trouble at all when energy is expressed in terms of momentum. What does Eq. 3.44 say about zero-mass particles such as photons? With $m = 0$, the second term of the square root becomes zero, and the square root of $P^2 c^2$ is just the magnitude of P times c. Why the *magnitude* of P? The left side of the equation is E, a real number. Therefore the right side must also be a real number. Putting this all together, we arrive at the simple equation

$$E = c|P|. \tag{3.45}$$

Energy for a massless particle is essentially the magnitude of the momentum vector, but for dimensional consistency, we multiply by the speed of light. Eq. 3.45 holds for photons. It's approximately true for neutrinos, which have a tiny mass. It does not hold for particles that move significantly slower than the speed of light.

3.5.3 An Example: Positronium Decay

Now that we know how to write down the energy of a massless particle, we can solve a simple but interesting problem. There's a particle called positronium that consists of an electron and a positron in orbit around each other. It's electrically neutral and its mass is approximately the mass of two electrons.[11]

---

[11] It's interesting to note that the mass of a positronium particle is slightly *less than* the sum of the masses of its constituent electron and positron. Why? Because it's bound. It has some kinetic energy due to the motion of its constituents, and that adds to its mass. But it has an even greater amount of negative potential energy. The negative potential energy outweighs the positive kinetic energy.

The positron is the electron's antiparticle, and if you let a positronium atom sit around for a while, these two anti-particles will annihilate each other, producing two photons. The positronium will disappear, and the two photons will go flying off in opposite directions. In other words, a neutral positronium particle with nonzero mass turns into pure electromagnetic energy. Can we calculate the energy and momentum of those two photons?

This would not make any sense at all in prerelativity physics. In prerelativity physics, the sum of the masses of particles is always unchanged. Chemical reactions happen, some chemicals turn into others, and so forth. But if you weigh the system—if you measure its mass—the sum of the ordinary masses never changes. However, when positronium decays into photons, the sum of the ordinary masses *does* change. The positronium particle has a finite nonzero mass, and the two photons that replace it are massless. The correct rule is *not* that the sum of masses is conserved; it's that the energy and the momentum are conserved. Let's consider momentum conservation first.

Suppose the positronium particle is at rest in your frame of reference.[12] Its momentum in this frame is zero, by definition. Now, the positronium atom decays to two photons. Our first conclusion is that the photons must go off back to back, in opposite directions, with equal and opposite momentum. If they don't travel in opposite directions, it's clear that the total momentum will not be zero. The final momentum must be zero because the initial momentum was zero. This means that the right-moving photon goes off with

---

[12] If not, just get moving and get into the frame of reference in which it *is* at rest.

momentum P, and the left-moving photon goes off with momentum $-P$.[13]

Now we can use the principle of energy conservation. Take your positronium atom, put it on a scale, and measure its mass. In its rest frame it has energy equal to mc^2. Because energy is conserved, this quantity must equal the combined energy of the two photons. Each of these photons must have the same energy as the other because their momenta have the same magnitude. Using Eq. 3.45, we can equate this energy to mc^2 as follows:

$$mc^2 = 2c|P|.$$

Solving for $|P|$, we find that

$$|P| = \frac{mc}{2}.$$

Each photon has a momentum whose absolute value is $mc/2$.

This is the mechanism by which mass turns into energy. Of course, the mass was *always* energy, but in the frozen form of rest energy. When the positronium atom decays, it results in two photons going out. The photons go on to collide with things. They may heat up the atmosphere; they could be absorbed by electrons, generate electrical currents, and so forth. The thing that's conserved is the total energy, not the individual masses of particles.

---

[13] We don't actually know what directions these back-to-back photons will take, except that they'll be moving in opposite directions. The line connecting the two photons is oriented randomly, according to the rules of quantum mechanics. However, there's nothing to stop us from orienting our x axis to coincide with their "chosen" direction of motion.

Lecture 4

Classical Field Theory

It's World Series time, and the bar in Hermann's Hideaway is packed with fans watching the game. Art arrives late and pulls up a stool next to Lenny.

Art: *What're the names of the guys in the outfield?*

Lenny: *Who's in classical field; What's in quantum field.*

Art: *Come on, Lenny, I just asked you. Who's in quantum field?*

Lenny: *Who's in classical field.*

Art: *That's what I'm asking you. All right, lemme try again. What about electric field?*

Lenny: *You want to know the name of the guy in electric field?*

Art: *Naturally.*

Lenny: *Naturally? No, Naturally's in magnetic field.*

So far, we have focused on the relativistic motion of particles. In this lecture, we'll introduce the theory of fields—not quantum field theory but relativistic classical field theory. There may be some occasional points of contact with quantum mechanics, and I'll point them out. But for the most part we'll stick to classical field theory.

The field theory that you probably know most about is the theory of electric and magnetic fields. These fields are vector quantities. They're characterized by a direction in space as well as a magnitude. We'll start with something a little easier: a scalar field. As you know, scalars are numbers that have magnitude but no direction. The field we'll consider here is similar to an important scalar field in particle physics. Perhaps you can guess which one it is as we go along.

4.1   Fields and Spacetime

Let's start with spacetime. Spacetime always has *one* time coordinate and some number of space coordinates. In principle, we could study physics with any number of time coordinates and any number of space coordinates. But in the physical world, even for theories in which spacetime may have ten dimensions, eleven dimensions, twenty-six dimensions, there's always exactly one time dimension. Nobody knows how to make logical sense out of more than one time dimension.

Let's call the space coordinates X^i and, for the moment, the time coordinate t. Keep in mind that in field theory the X^i are not degrees of freedom; they are merely labels that label the points of space. The events of spacetime are labeled (t, X^i). The index i runs over as many space coordinates as there are.

Not surprisingly, the degrees of freedom of field theory are fields. A field is a measurable quantity that depends on position in space and may also vary with time. There are plenty of examples from common, garden-variety physics. Atmospheric temperature varies from place to place and time to time. A possible notation for it would be $T(t, X^i)$. Because it has only one component—a single number—it's a *scalar* field. Wind velocity is a *vector* field because velocity has a direction, which may also vary over space and time.

Mathematically, we represent a field as a function of space and time. We often label this function with the Greek letter ϕ:

$$\phi(t, X^i).$$

It's common in field theory to say that spacetime is $(3 + 1)$-dimensional, meaning that there are three dimensions of space and one dimension of time. More generally we might be interested in studying fields in spacetimes with other numbers of space dimensions. If a spacetime has d space dimensions, we would call it $(d + 1)$-dimensional.

4.2   Fields and Action

As I mentioned earlier, the principle of least action is one of the most fundamental principles of physics, governing all known laws of physics. Without it we would have no reason to believe in energy conservation or even the existence of solutions to the equations we write down. We will also base our study of fields on an action principle. The action principles that govern fields are generalizations of those for particles. Our plan is to examine in parallel the action principles that govern fields and those that govern particles, comparing them as we go. To simplify this comparison, we will first restate the action principle for nonrelativistic particles in the language of fields.

4.2.1   Nonrelativistic Particles Redux

I want to briefly go back to the theory of nonrelativistic particles, not because I am really interested in slow particles, but because the mathematics has some similarity with the theory of fields. In fact, in a certain formal sense it is a field theory of a simple kind—a field theory in a world whose spacetime has *zero* space dimensions, and as always, one time dimension.

To see how this works, let's consider a particle that moves along the x axis. Ordinarily we would describe the motion of the particle by a trajectory $x(t)$. However, with no change in the content of the theory, we might change the *notation* and call the position of the particle ϕ. Instead of $x(t)$, the trajectory would be described by $\phi(t)$.

If we were to re-wire the meaning of the symbol $\phi(t)$, that is, if we use it to represent a scalar field, it would become a special case of $\phi(t, X^i)$—a special case in which there are *no* dimensions of space. In other words, a particle theory in one dimension of space has the same mathematical structure as a scalar field theory in zero dimensions of space. Physicists sometimes refer to the theory of a single particle as a field theory in $(0 + 1)$-dimensions, the one dimension being time.

Fig. 4.1 illustrates the motion of a nonrelativistic particle. Notice that we use the horizontal axis for time, just to emphasize that t is the independent parameter. On the vertical axis we plot the position of the particle at time t, calling it $\phi(t)$. The curve $\phi(t)$ represents the history of the particle's motion. It tells you what the position ϕ is at each moment of time. As the diagram shows, ϕ can be negative or positive. We characterize this trajectory using the principle of least action.

As you recall, action is defined as the integral of some

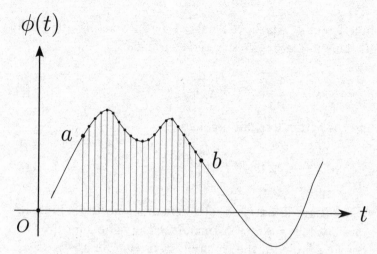

Figure 4.1: Nonrelativistic Particle Trajectory.

Lagrangian $\mathcal{L}$, from an initial time a to a final time b:

$$Action = \int_a^b \mathcal{L}\,dt.$$

For nonrelativistic particles, the Lagrangian is simple; it's the kinetic energy *minus* the potential energy. Kinetic energy is usually expressed as $\frac{1}{2}mv^2$, but in our new notation we would write $\dot{\phi}$ or $\dfrac{d\phi}{dt}$ instead of v for velocity. With this notation, the kinetic energy becomes $\frac{1}{2}m\dot{\phi}^2$, or $\frac{1}{2}m\left(\dfrac{d\phi}{dt}\right)^2$. We'll simplify things a little by setting the mass m equal to 1. Thus the kinetic energy is

$$Kinetic\ Energy = \frac{1}{2}\left(\frac{d\phi}{dt}\right)^2.$$

What about potential energy? In our example, potential energy is just a function of position—in other words it's a function of ϕ, which we'll call $V(\phi)$. Subtracting $V(\phi)$ from the kinetic energy gives us the Lagrangian

$$\mathcal{L} = \frac{1}{2}\left(\frac{d\phi}{dt}\right)^2 - V(\phi), \tag{4.1}$$

and the action integral becomes

$$Action = \int_a^b \left[\frac{1}{2}\left(\frac{d\phi}{dt}\right)^2 - V(\phi)\right] dt. \tag{4.2}$$

As we know from classical mechanics, the Euler-Lagrange equation tells us how to minimize the action integral and therefore provides the equation of motion for the particle.[1] For this example, the Euler-Lagrange equation is

$$\frac{d}{dt}\frac{\partial \mathcal{L}}{\partial \dot{\phi}} = \frac{\partial \mathcal{L}}{\partial \phi},$$

and our task is to apply this equation to the Lagrangian of Eq. 4.1. Let's start by writing down the derivative of $\mathcal{L}$ with respect to $\left(\dfrac{d\phi}{dt}\right)$:

$$\frac{\partial \mathcal{L}}{\partial\left(\dfrac{d\phi}{dt}\right)} = \frac{d\phi}{dt}.$$

Next, the Euler-Lagrange equation instructs us to take the time derivative of this result:

$$\frac{d}{dt}\frac{\partial \mathcal{L}}{\partial\left(\dfrac{d\phi}{dt}\right)} = \frac{d^2\phi}{dt^2}.$$

This completes the left side of the Euler-Lagrange equation.

---

[1] There's only one Euler-Lagrange equation in this example because the only variables are ϕ and $\dot{\phi}$.

Now for the right side. Referring once more to Eq. 4.1, we find that

$$\frac{\partial \mathcal{L}}{\partial \phi} = -\frac{\partial V(\phi)}{\partial \phi}.$$

Finally, setting the left side equal to the right side gives us

$$\frac{d^2\phi}{dt^2} = -\frac{\partial V(\phi)}{\partial \phi}. \qquad (4.3)$$

This equation should be familiar. It's just Newton's equation for the motion of a particle. The right side is force, and the left side is acceleration. This would be Newton's second law $F = ma$ had we not set the mass to 1.

The Euler-Lagrange equations provide the solution to the problem of finding the trajectory that a particle follows between two fixed points a and b. They're equivalent to finding the trajectory of least action that connects the two fixed end points.[2]

As you know, there's another way to think of this. You can divide the time axis into a lot of little pieces by drawing lots of closely spaced vertical lines on Fig. 4.1. Instead of thinking of the action as an integral, just think of it as a sum of terms. What do those terms depend on? They depend on the value of $\phi(t)$ and its derivatives at each time. In other words, the total action is simply a function of many values of $\phi(t)$. How do you minimize a function of ϕ? You differentiate with respect to ϕ. That's what the Euler-Lagrange equations accomplish. Another way to say it is that they're the solution to the problem of moving these points around until you find the trajectory that minimizes the action.

---

[2] I often say *least* or *minimum*, but you know very well that I really mean *stationary*: The action could be a maximum *or* a minimum.

4.3   Principles of Field Theory

Thus far we've been studying a field theory in a world with no space dimensions. Based on this example, let's try to develop some intuition about a theory for a world more like the one we live in—a world with one or more space dimensions. We take it as a given that field theory, in fact the whole world, is governed by an action principle. Stationary action is a powerful principle that encodes and summarizes a huge number of physics laws.

4.3.1   The Action Principle

Let's define the action principle for fields. For a particle moving in one dimension (Fig. 4.1) we chose two fixed end points, a and b, as boundaries along the time axis. Then we considered all possible trajectories that connect the two boundary points and asked for the particular curve that minimizes the action. (This is a lot like finding the shortest distance between two points.) In this way the principle of least action tells us how to fill in the value of $\phi(t)$ between the boundary points a and b.

The problem of field theory is a generalization of this idea of filling in the boundary data. We begin with a spacetime region which we may visualize as a four-dimensional box. To construct it we take a three-dimensional box of space—let's say the space inside some cube—and consider it for some interval of time. That forms a four-dimensional box of spacetime. In Fig. 4.2 I've tried to illustrate such a spacetime box but with only two space directions.

The general problem of field theory can be stated in the following way: Given the values of ϕ everywhere on the boundary of the spacetime box, determine the field everywhere inside the box. The rules of the game are similar to the

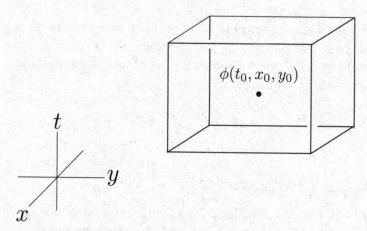

Figure 4.2: Boundary of a Spacetime Region for Applying the Principle of Least Action. Only two space dimensions are shown.

particle case. We will need an expression for the action—we'll come to that soon—but let's assume we know the action for every configuration of the field in the box. The principle of least action tells us to wiggle the field until we find the particular function $\phi(t, x, y, z)$ that gives the least action.

For particle motion the action was constructed by adding up little bits of action for each infinitesimal time segment. This gave an integral over the time interval between the boundary points a and b in Fig. 4.1. The natural generalization in field theory is to construct an action as a sum over tiny spacetime cells—in other words, an integral over the spacetime box in Fig. 4.2.

$$Action = \int \mathcal{L} \; dt \; dx \; dy \; dz,$$

where $\mathcal{L}$ is a Lagrangian that we still have not specified. But in relativity we've learned to think of these four coordinates on an equal footing, with each of them being part of spacetime.

We blur the distinction between space and time by giving them similar names—we just call them X^μ—where the index μ runs over all four coordinates, and it's standard practice to write the preceding integral as

$$Action = \int \mathcal{L} d^4x.$$

4.3.2   Stationary Action for ϕ

Because the Lagrangian $\mathcal{L}$ for a field is integrated over space as well as time, it's often called the Lagrange *density*.[3]

What variables does $\mathcal{L}$ depend on? Return for a moment to the nonrelativistic particle. The Lagrangian depends on the coordinates of the particle, and the velocities. In the notation we've been using, the Lagrangian depends on ϕ and $\left(\dfrac{\partial \phi}{\partial t}\right)$. The natural generalization, inspired by Minkowski's idea of spacetime, is for $\mathcal{L}$ to depend on ϕ and the partial derivatives of ϕ with respect to *all* coordinates. In other words, $\mathcal{L}$ depends on

$$\phi, \; \frac{\partial \phi}{\partial t}, \; \frac{\partial \phi}{\partial x}, \; \frac{\partial \phi}{\partial y}, \; \frac{\partial \phi}{\partial z}.$$

Thus we write

$$Action = \int \mathcal{L}\left(\phi, \frac{\partial \phi}{\partial X^\mu}\right) d^4x,$$

where the index μ ranges over the time coordinate and all

---

[3] This is just a matter of dimensional consistency. Unlike particle motion, the action integral for a field is taken over $dxdydz$, as well as dt. For the action to have the same units for fields as it does for particles, the Lagrangian must carry units of energy divided by volume. Hence the term *density*.

three space coordinates. In this action integral, we did not write $\mathcal{L}$ with explicit dependencies on t, x, y, or z. But for some problems, $\mathcal{L}$ could indeed depend on these variables, just as the Lagrangian for particle motion might depend explicitly on time. For a closed system—a system where energy and momentum are conserved—$\mathcal{L}$ does not depend explicitly on time or spatial position.

As in ordinary classical mechanics, the equations of motion are derived by wiggling ϕ and minimizing the action. In *Volume I*, I showed that the result of this wiggling process can be achieved by applying a special set of equations called the Euler-Lagrange equations. For the motion of a particle, the Euler-Lagrange equations take the form

$$\frac{d}{dt} \frac{\partial \mathcal{L}}{\partial \left(\dfrac{\partial \phi}{\partial t} \right)} - \frac{\partial \mathcal{L}}{\partial \phi} = 0. \tag{4.4}$$

How would the Euler-Lagrange equations change for the multidimensional spacetime case? Let's look closely at each term on the left side. Clearly, we need to modify Eq. 4.4 to incorporate all four directions of spacetime. The first term, which already references the time direction, becomes a sum of terms, one for each direction of spacetime. The correct prescription is to replace Eq. 4.4 with

$$\sum_{\mu} \frac{\partial}{\partial X^{\mu}} \frac{\partial \mathcal{L}}{\partial \left(\dfrac{\partial \phi}{\partial X^{\mu}} \right)} - \frac{\partial \mathcal{L}}{\partial \phi} = 0. \tag{4.5}$$

The first term in the sum is just

$$\frac{\partial}{\partial t} \frac{\partial \mathcal{L}}{\partial \left(\dfrac{\partial \phi}{\partial t} \right)},$$

whose similarity to the first term in Eq. 4.4 is obvious. The other three terms involve analogous spatial derivatives that fill out the expression and give it its spacetime character.

Eq. 4.5 is the Euler-Lagrange equation for a single scalar field. As we will see, these equations are closely related to wave equations that describe wavelike oscillations of ϕ.

As is true in particle mechanics, there may be more than a single degree of freedom. In the case of field theory, that would mean more than one field. Let's be explicit and suppose there are two fields ϕ and χ. The action will depend on both fields and their derivatives,

$$\phi, \; \frac{\partial \phi}{\partial X^\mu}$$

$$\chi, \; \frac{\partial \chi}{\partial X^\mu}.$$

If there are two fields, then there must be an Euler-Lagrange equation for each field:

$$\sum_\mu \frac{\partial}{\partial X^\mu} \frac{\partial \mathcal{L}}{\partial \left(\dfrac{\partial \phi}{\partial X^\mu} \right)} - \frac{\partial \mathcal{L}}{\partial \phi} = 0$$

$$\sum_\mu \frac{\partial}{\partial X^\mu} \frac{\partial \mathcal{L}}{\partial \left(\dfrac{\partial \chi}{\partial X^\mu} \right)} - \frac{\partial \mathcal{L}}{\partial \chi} = 0. \qquad (4.6)$$

More generally, if there are several fields, there will be an Euler-Lagrange equation for every one of them.

Incidentally, while it's true that we're developing this example as a *scalar* field, so far we haven't said anything that *requires* ϕ to be a scalar. The only hint of ϕ's scalar character is the fact that we're working in only one component; a vector

field would have additional components, and each new component would generate an additional Euler-Lagrange equation.

4.3.3 More About Euler-Lagrange Equations

The Lagrangian in Eq. 4.1 for a nonrelativistic particle
contains a kinetic energy term proportional to

$$\frac{1}{2}\left(\frac{d\phi}{dt}\right)^2$$

and a potential energy term

$$-V(\phi).$$

We might guess that the generalization to a field theory
would look similar, but with the kinetic energy term having
space as well as time derivatives:

$$\mathcal{L} = \frac{1}{2}\left[\left(\frac{\partial\phi}{\partial t}\right)^2 + \left(\frac{\partial\phi}{\partial x}\right)^2 + \left(\frac{\partial\phi}{\partial y}\right)^2 + \left(\frac{\partial\phi}{\partial z}\right)^2\right] - V(\phi).$$

But there is something obviously wrong with this guess:
Space and time enter into it in exactly the same way. Even in
relativity theory, time is not completely symmetric with
space, and ordinary experience tells us they are different.
One hint, coming from relativity, is the sign difference
between space and time in the expression for proper time,

$$d\tau^2 = dt^2 - dx^2 - dy^2 - dz^2.$$

Later we will see that the derivatives $\frac{\partial\phi}{\partial X^\mu}$ form the components of a 4-vector and that

$$\left(\frac{\partial\phi}{\partial t}\right)^2 - \left(\frac{\partial\phi}{\partial x}\right)^2 - \left(\frac{\partial\phi}{\partial y}\right)^2 - \left(\frac{\partial\phi}{\partial z}\right)^2$$

defines a Lorentz invariant quantity. This suggests that we replace the sum of squares (of derivatives) with the difference of squares. Later, when we bring Lorentz invariance back into the picture, we'll see why this makes sense. We'll take this modified Lagrangian,

$$\mathcal{L} = \frac{1}{2}\left[\left(\frac{\partial\phi}{\partial t}\right)^2 - \left(\frac{\partial\phi}{\partial x}\right)^2 - \left(\frac{\partial\phi}{\partial y}\right)^2 - \left(\frac{\partial\phi}{\partial z}\right)^2\right] - V(\phi),$$

$$(4.7)$$

to be the Lagrangian for our field theory. We can also view it as a prototype. Later on, we'll build other theories out of Lagrangians that are similar to this one.

The function $V(\phi)$ is called the *field potential*. It is analogous to the potential energy of a particle. More precisely, it is an energy density (energy per unit volume) at each point of space that depends on the value of the field at that point. The function $V(\phi)$ depends on the context, and in at least one important case it is deduced from experiment. We'll discuss that case in Section 4.5.1. For the moment, $V(\phi)$ can be any function of ϕ.

Let's take this Lagrangian and work out its equations of motion step by step. The Euler-Lagrange equation in Eq. 4.5 tells us exactly how to proceed. We allow the index μ in this equation to take on the values 0, 1, 2, and 3. Each of these index values generates a separate term in the resulting differential equation for the field. Here's how it works:

Step 1. We start by setting the index μ equal to zero. In other words, X^μ starts out as X^0, which corresponds to the time coordinate. Calculating the derivative of $\mathcal{L}$ with respect to $\frac{\partial\phi}{\partial t}$ gives us the simple result

$$\frac{\partial \mathcal{L}}{\partial \left(\dfrac{\partial \phi}{\partial t} \right)} = \frac{\partial \phi}{\partial t}.$$

Eq. 4.5 now instructs us to take one more derivative:

$$\frac{\partial}{\partial X^0} \frac{\partial \mathcal{L}}{\partial \left(\dfrac{\partial \phi}{\partial t} \right)} = \frac{\partial^2 \phi}{\partial t^2}.$$

This term, $\dfrac{\partial^2 \phi}{\partial t^2}$, is analogous to the acceleration of a particle.

Step 2. Next, we set index μ equal to 1, and X^μ now becomes X^1, which is just the x coordinate. Calculating the derivative of $\mathcal{L}$ with respect to $\dfrac{\partial \phi}{\partial x}$ gives us

$$\frac{\partial \mathcal{L}}{\partial \left(\dfrac{\partial \phi}{\partial x} \right)} = -\frac{\partial \phi}{\partial x}$$

and

$$\frac{\partial}{\partial x} \frac{\partial \mathcal{L}}{\partial \left(\dfrac{\partial \phi}{\partial x} \right)} = \frac{\partial}{\partial x} \left(-\frac{\partial \phi}{\partial x} \right) = -\frac{\partial^2 \phi}{\partial x^2}.$$

We get similar results for the y and z coordinates when we set μ equal to 2 and 3 respectively.

Step 3. Rewrite Eq. 4.5 using the previous two steps:

$$\frac{\partial^2 \phi}{\partial t^2} - \frac{\partial^2 \phi}{\partial x^2} - \frac{\partial^2 \phi}{\partial y^2} - \frac{\partial^2 \phi}{\partial z^2} + \frac{\partial V}{\partial \phi} = 0. \qquad (4.8)$$

As usual, if we want to use conventional units we need to restore the factors of c. This is easily done and the equation of motion becomes

$$\frac{1}{c^2}\frac{\partial^2 \phi}{\partial t^2} - \frac{\partial^2 \phi}{\partial x^2} - \frac{\partial^2 \phi}{\partial y^2} - \frac{\partial^2 \phi}{\partial z^2} + \frac{\partial V}{\partial \phi} = 0. \qquad (4.9)$$

4.3.4   Waves and Wave Equations

Of all the phenomena described by classical field theory, the most common and easy to understand is the propagation of waves. Sound waves, light waves, water waves, waves along vibrating guitar strings: All are described by similar equations, which are unsurprisingly called wave equations. This connection between field theory and wave motion is one of the most important in physics. It's now time to explore it.

Eq. 4.9 is a generalization of Newton's equation of motion

$$\frac{d^2 \phi}{dt^2} = -\frac{\partial V(\phi)}{\partial \phi}$$

(Eq. 4.3) for fields. The term $\dfrac{\partial V(\phi)}{\partial \phi}$ represents a kind of force acting on the field, pushing it away from some natural unforced motion. For a particle, the unforced motion is uniform motion with a constant velocity. For fields, the unforced motion is propagating waves similar to sound or electromagnetic waves. To see this, let's do two things to simplify Eq. 4.9. First we can drop the force term by setting $V(\phi) = 0$. Second, instead of three dimensions of space we will study the equation with only one space direction, x. Eq. 4.9 now takes the much simpler form of a $(1 + 1)$-dimensional wave equation,

$$\frac{1}{c^2}\frac{\partial^2 \phi}{\partial t^2} - \frac{\partial^2 \phi}{\partial x^2} = 0. \qquad (4.10)$$

I will show you a whole family of solutions. Let's consider

any function of the combination $(x + ct)$. We'll call it $F(x + ct)$. Now consider its derivatives with respect to x and t. It is easy to see that these derivatives satisfy

$$\frac{\partial F(x + ct)}{\partial t} = c \frac{\partial F(x + ct)}{\partial x}.$$

If we apply the same rule a second time we get

$$\frac{\partial^2 F(x + ct)}{\partial t^2} = c^2 \frac{\partial^2 F(x + ct)}{\partial x^2}$$

or

$$\frac{1}{c^2} \frac{\partial^2 F(x + ct)}{\partial t^2} - \frac{\partial^2 F(x + ct)}{\partial x^2} = 0. \qquad (4.11)$$

Eq. 4.11 is nothing but the wave equation of Eq. 4.10 for the function F. We have found a large class of solutions to the wave equation. Any function of the combination $(x + ct)$ is such a solution.

What are the properties of a function like $F(x + ct)$? At time $t = 0$ it is just the function $F(x)$. As time goes on, the function changes, but in a simple way: It moves rigidly to the left (in the negative x direction) with velocity c. Let's take a particular example in which F (which we now identify with ϕ) is a sine function with wave number k:

$$\phi(t, x) = \sin k(x + ct).$$

This is a left-moving sine wave, moving with velocity c. There are also cosine solutions as well as wave packets and pulses. As long as they move rigidly to the left with velocity c they are solutions of the wave equation, Eq. 4.10.

What about right-moving waves? Are they also described by Eq. 4.10? The answer is, yes they are. All you have to do to change $F(x + ct)$ to a right-moving wave is replace $x + ct$ with $x - ct$. I will leave it to you to show that any function of the form $F(x - ct)$ moves rigidly to the right and that it also satisfies Eq. 4.10.

4.4   Relativistic Fields

When we built our theory of particles, we used two principles in addition to the stationary action principle. First, the action is always an integral. For particles, it's an integral along a trajectory. We've already dealt with this issue for fields by redefining the action as an integral over spacetime. Second, the action needs to be invariant; it should be built up from quantities in such a way that it has the exact same form in every reference frame.

How did we manage that for the case of a particle? We built up the action by slicing the trajectory into lots of little segments, calculating the action for each segment, and then adding all these small action elements together. Then, we took the limit of this process, where the size of each segment approaches zero and the sum becomes an integral. The crucial point is this: When we defined the action for one of the segments, we chose a quantity that was invariant—the proper time along the trajectory. Because all observers agree on the value of the proper time for each little segment, they will agree about the number you get when you add them all together. As a result, the equations of motion you derive using the stationary action principle have exactly the same form in every reference frame. The laws of particle mechanics are invariant. I already hinted on how to do this for fields in Section 4.3.3, but now we want to get serious about Lorentz invariance in field theory.

We will need to know how to form invariant quantities from fields, and then use them to construct action integrals that are also invariant. To do so, we'll need a clear concept of the transformation properties of fields.

4.4.1 Field Transformation Properties

Let's get back to Art in the railway station and Lenny moving past him in the train. They both look at the same event—Art calls it (t, x, y, z) and Lenny calls it (t', x', y', z'). Their special field detectors register numerical values for some field ϕ. The simplest possible kind of field is one for which they obtain exactly the same result. If we call the field that Art measures $\phi(t, x, y, z)$ and the field that Lenny measures $\phi'(t', x', y', z')$, then the simplest transformation law would be

$$\phi'(t', x', y', z') = \phi(t, x, y, z). \tag{4.12}$$

In other words, at any particular point of spacetime Art and Lenny (and everyone else) agree about the value of the field ϕ at that point.

A field with this property is called a *scalar field*. The idea of a scalar field is illustrated in Fig. 4.3. It's important to understand that coordinates (t', x') and (t, x) both reference the same point in spacetime. Fig. 4.3 drives this point home. The unprimed coordinates represent spacetime position in Art's frame. The primed axes stand for spacetime position in Lenny's frame. The label $\phi(t, x)$ is Art's name for the field; Lenny calls the same thing $\phi'(t', x')$ since he is the primed observer. But it's the same field, with the same value at the same point in spacetime. The value of a scalar field at a spacetime point is invariant.

Not all fields are scalars. Here's an example from ordinary nonrelativistic physics: wind velocity. Wind velocity is a vector with three components. Observers whose axes are oriented differently will not agree on the values of the components. Art and Lenny will certainly not agree. The air might be still in Art's frame, but when Lenny sticks his head out the window he detects a large wind velocity.

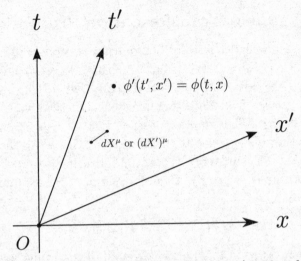

Figure 4.3: Transformations. Fields ϕ and ϕ' both reference the same point in spacetime. Both fields have the same value at this point. Displacements dX^μ and $(dX')^\mu$ both reference the same spacetime interval, but the primed components are different from the unprimed components. They're related by Lorentz transformation.

Next we come to 4-vector fields. A good example is one I've already mentioned: wind velocity. Here is how we might define it in Art's frame. At the spacetime point (t, x, y, z), Art measures the local components of the velocity of the air molecules dX^i/dt and maps them out. The result is a 3-vector field with components

$$V^x(t, x, y, z),\ \ V^y(t, x, y, z),\ \ V^z(t, x, y, z).$$

But you might already guess that in a relativistic theory we should represent the molecular velocities relativistically: not by $V^i = dX^i/dt$ but by $U^\mu = dX^\mu/d\tau$. Mapping out the relativistic wind velocity

$$U^\mu(t, x, y, z)$$

or

$$U^\mu(X^0, X^1, X^2, X^3)$$

would define a 4-vector field.

Given Art's description of the relativistic wind velocity, we can ask: What are its components in Lenny's frame? Since 4-velocity is a 4-vector, the answer follows from the Lorentz transformation that we wrote down earlier in Eq. 2.15:

$$(U')^0 = \frac{U^0 - vU^1}{\sqrt{1 - v^2}}$$

$$(U')^1 = \frac{U^1 - vU^0}{\sqrt{1 - v^2}}$$

$$(U')^2 = U^2$$

$$(U')^3 = U^3. \tag{4.13}$$

I didn't include the dependence on the coordinates in Eq. 4.13 as it would have overly cluttered the equations, but the rule is simple: The U on the right side are functions of (t, x, y, z) and the U' on the left side are functions of (t', x', y', z'), but both sets of coordinates refer to the same spacetime point.

Let's consider another complex of four quantities that we'll call the spacetime gradient of the scalar field ϕ. Its components are the derivatives of ϕ. We'll use the shorthand notation $\partial_\mu \phi$, defined by

$$\partial_\mu \phi = \frac{\partial \phi}{\partial X^\mu}. \tag{4.14}$$

For example,

$$\partial_1 \phi = \frac{\partial \phi}{\partial X^1}.$$

In a slightly different notation, we can write

$$\partial_t \phi = \frac{\partial \phi}{\partial t}$$

$$\partial_x \phi = \frac{\partial \phi}{\partial x}$$

$$\partial_y \phi = \frac{\partial \phi}{\partial y}$$

$$\partial_z \phi = \frac{\partial \phi}{\partial z}.$$

We might expect that ϕ_μ is a 4-vector and transforms the same way as U^μ; we would be mistaken, although not by a lot.

4.4.2 Mathematical Interlude: Covariant Components

I want to pause here for a couple of mathematical points about transformations from one set of coordinates to another. Let's suppose we have a space described by two sets of coordinates X^μ and $(X')^\mu$. These could be Art's and Lenny's spacetime coordinates, but they needn't be. Let's also consider an infinitesimal interval described by dX^μ or $d(X')^\mu$. Ordinary multivariable calculus implies the following relation between the two sets of differentials:

$$d(X')^\mu = \sum_\nu \frac{\partial(X')^\mu}{\partial X^\nu} \, dX^\nu. \qquad (4.15)$$

Einstein wrote many equations of this form. After a while he noticed a pattern: Whenever he had a repeated index in a single expression—the index ν on the right side of the equation is such a repeated index—it was always summed over. In one of his papers on general relativity, after several pages, he apparently got tired of writing the summation sign and simply said that from now on, whenever an expression had a repeated index he would assume that it was summed over. That convention became known as the *Einstein summation convention*. It is completely ubiquitous today, to the point where no one in physics even bothers to mention it. I'm also tired of writing $\sum_\nu$, so from now on we will use Einstein's clever convention, and Eq. 4.15 becomes

$$d(X')^\mu = \frac{\partial(X')^\mu}{\partial X^\nu} \, dX^\nu. \qquad (4.16)$$

If the equations relating X and X' are linear, as they would be for Lorentz transformations, then the partial derivatives $\dfrac{\partial(X')^\mu}{\partial X^\nu}$ are constant coefficients. Let's take the Lorentz transformations

$$(X')^0 = \frac{X^0 - vX^1}{\sqrt{1 - v^2}}$$

$$(X')^1 = \frac{X^1 - vX^0}{\sqrt{1 - v^2}}, \qquad (4.17)$$

as an example. Here is a list of the four constant coefficients that would result from Eq. 4.16:

$$\frac{\partial (X')^1}{\partial X^1} = \frac{1}{\sqrt{1 - v^2}}$$

$$\frac{\partial (X')^1}{\partial X^0} = \frac{-v}{\sqrt{1 - v^2}}$$

$$\frac{\partial (X')^0}{\partial X^1} = \frac{-v}{\sqrt{1 - v^2}}$$

$$\frac{\partial (X')^0}{\partial X^0} = \frac{1}{\sqrt{1 - v^2}}. \tag{4.18}$$

If we plug these into Eq. 4.16 we get the expected result,

$$d(X')^0 = \frac{dX^0}{\sqrt{1 - v^2}} - \frac{v dX^1}{\sqrt{1 - v^2}}$$

$$d(X')^1 = \frac{dX^1}{\sqrt{1 - v^2}} - \frac{v dX^0}{\sqrt{1 - v^2}}.$$

This, of course, is the perfectly ordinary Lorentz transformation of the components of a 4-vector.

Let's abstract from this exercise a general rule for the transformation of 4-vectors. Going back to Eq. 4.16, let's replace 4-vector components $d(X')^\mu$ and dX^ν with $(A')^\mu$ and A^ν. These represent the components of *any* 4-vector A in frames related by a coordinate transformation. The generalization of Eq. 4.16 becomes

$$(A')^\mu = \frac{\partial (X')^\mu}{\partial X^\nu} A^\nu, \tag{4.19}$$

where ν is a summation index. Eq. 4.19 is the general rule for transforming the components of a 4-vector. For the important special case of a Lorentz transformation, this becomes

$$(A')^0 = \frac{A^0}{\sqrt{1-v^2}} - \frac{vA^1}{\sqrt{1-v^2}}$$

$$(A')^1 = \frac{A^1}{\sqrt{1-v^2}} - \frac{vA^0}{\sqrt{1-v^2}}. \qquad (4.20)$$

My real reason for doing this was not to explain how dX^μ or A^μ transforms, but to set up the calculation for transforming $\partial_\mu \phi$. These objects—there are four of them—also form the components of a 4-vector, although of a slightly different kind than the dX^μ. They obviously refer to the coordinate system X but can be transformed to the X' frame.

The basic transformation rule derives from calculus, and it's a multivariable generalization of the chain rule for derivatives. Let me remind you of the ordinary chain rule. Let $\phi(x)$ be a function of the coordinate X, and let the primed coordinates X' also be a function of X. The derivative of ϕ with respect to X' is given by the chain rule,

$$\frac{\partial \phi}{\partial X'} = \frac{\partial X}{\partial X'} \frac{\partial \phi}{\partial X}.$$

The multivariable generalization involves a field ϕ that depends on several independent coordinates X^μ and a second set of coordinates $(X')^\nu$. The generalized chain rule reads

$$\frac{\partial \phi}{\partial (X')^\nu} = \sum_\mu \frac{\partial X^\mu}{\partial (X')^\nu} \frac{\partial \phi}{\partial X^\mu},$$

or using the summation convention, and the shorthand

notation in Eq. 4.14,

$$\partial'_\nu \phi = \frac{\partial X^\mu}{\partial (X')^\nu} \, \partial_\mu \phi. \tag{4.21}$$

Let's be more general and replace $\partial_\mu \phi$ with A_μ so that Eq. 4.21 becomes

$$A'_\nu = \frac{\partial X^\mu}{\partial (X')^\nu} \, A_\mu. \tag{4.22}$$

Take a moment to compare Eqs. 4.19 and 4.22. I'll write them again to make it easy to compare—first Eq. 4.19 and then Eq. 4.22:

$$(A')^\mu = \frac{\partial (X')^\mu}{\partial X^\nu} A^\nu$$

$$A'_\nu = \frac{\partial X^\mu}{\partial (X')^\nu} \, A_\mu.$$

There are two differences. The first is that in Eq. 4.19 the Greek indices on A appear as superscripts while in Eq. 4.22 they appear as subscripts. That hardly seems important, but it is. The second difference is the coefficients: In Eq. 4.19 they are derivatives of X' with respect to X, while in Eq. 4.22 they are derivatives of X with respect to X'.

Evidently there are two different kinds of 4-vectors, which transform in different ways: the kind that have superscripts and the kind with subscripts. They are called *contravariant components* (superscripts) and *covariant components* (subscripts), but since I always forget which is which, I just call them upper and lower 4-vectors. Thus the 4-vector dX^μ is a contravariant or upper 4-vector, while the spacetime gradient $\partial_\mu \phi$ is a covariant or lower 4-vector.

Let's go back to Lorentz transformations. In Eq. 4.18, I

wrote down the coefficients for the transformation of upper 4-vectors. Here they are again:

$$\frac{\partial (X')^1}{\partial X^1} = \frac{1}{\sqrt{1 - v^2}}$$

$$\frac{\partial (X')^1}{\partial X^0} = \frac{-v}{\sqrt{1 - v^2}}$$

$$\frac{\partial (X')^0}{\partial X^1} = \frac{-v}{\sqrt{1 - v^2}}$$

$$\frac{\partial (X')^0}{\partial X^0} = \frac{1}{\sqrt{1 - v^2}}$$

We can make a similar list for the coefficients in Eq. 4.22 for lower 4-vectors. In fact, we don't have much work to do. We can get them by interchanging the primed and unprimed coordinates, which for Lorentz transformations just means interchanging the rest and moving frames. This is especially easy since it only requires us to change the sign of the velocity (remember, if Lenny moves with velocity v in Art's frame, then Art moves with velocity $-v$ in Lenny's frame). All we have to do is to interchange primed and unprimed coordinates and at the same time reverse the sign of v.

$$\frac{\partial X^1}{\partial (X')^1} = \frac{1}{\sqrt{1 - v^2}}$$

$$\frac{\partial X^1}{\partial (X')^0} = \frac{v}{\sqrt{1 - v^2}}$$

$$\frac{\partial X^0}{\partial (X')^1} = \frac{v}{\sqrt{1 - v^2}}$$

$$\frac{\partial X^0}{\partial (X')^0} = \frac{1}{\sqrt{1 - v^2}}.$$

Here then are the transformation rules for the components of covariant (lower) 4-vectors:

$$(A')_0 = \frac{A_0}{\sqrt{1 - v^2}} + \frac{vA_1}{\sqrt{1 - v^2}}$$

$$(A')_1 = \frac{A_1}{\sqrt{1 - v^2}} + \frac{vA_0}{\sqrt{1 - v^2}}. \qquad (4.23)$$

We've already seen examples such as X^μ, displacement from the origin. The differential displacement between neighboring points, dX^μ, is also a 4-vector. If you multiply 4-vectors by scalars (that is, by invariants), the result is also a 4-vector. That's because invariants are completely passive when you transform. We've already seen that the proper time $d\tau$ is invariant, and therefore the quantity $dX^\mu/d\tau$, which we call 4-velocity, is also a 4-vector:

$$U^\mu = \frac{dX^\mu}{d\tau}.$$

When we say "U^μ is a 4-vector," what do we actually mean? We mean that its behavior in other reference frames is governed by the Lorentz transformation. Let's recall the Lorentz transformation for the coordinates of two reference frames whose relative velocity is v along the x axis:

$$t' = \frac{t - vx}{\sqrt{1 - v^2}}$$

$$x' = \frac{x - vt}{\sqrt{1 - v^2}}$$

$$y' = y$$

$$z' = z.$$

If a complex of four quantities (consisting of a time component and three space components) transforms in this way, we call it a 4-vector. As you know, differential displacements also have this property:

$$dt' = \frac{dt - vdx}{\sqrt{1 - v^2}}$$

$$dx' = \frac{dx - vdt}{\sqrt{1 - v^2}}$$

$$dy' = dy$$

$$dz' = dz.$$

Table 4.1 summarizes the transformation properties of scalars and 4-vectors. We use the slightly abstract notation A^μ to represent an arbitrary 4-vector. A^0 is the time component, and each of the other components represents a direction in space.

An example of a field with these properties would be a fluid that fills all of spacetime. At every point in the fluid there would be a 4-velocity as well as an ordinary 3-velocity. We could call this 4-velocity $U^\mu(t, x)$. If the fluid flows, the velocity might be different in different places. The 4-velocity of such a fluid can be thought of as a field. Because it's a 4-velocity, it's automatically a 4-vector and would transform in exactly the same way as our prototype 4-vector A^μ. The

Field	Transformation	Examples
Scalars:	Same Value	Temperature
	$\phi'(t', x') = \phi(t, x)$	Proper Time
Vectors:	Lorentz Transform	Displacement: X^μ, dX^μ
	$(A')^0 = \dfrac{A^0 - vA^1}{\sqrt{1 - v^2}}$	Any 4-Vector: A^μ
	$(A')^1 = \dfrac{A^1 - vA^0}{\sqrt{1 - v^2}}$	
	$(A')^2 = A^2$	
	$(A')^3 = A^3$	

Table 4.1: Field Transformations. Greek index μ takes values 0, 1, 2, 3, which correspond to t, x, y, z in ordinary (3+1)-dimensional spacetime. In nonrelativistic physics, ordinary Euclidean distance is also considered a scalar.

values of U's components in your frame would be different from their values in my frame; they would be related by the equations in Table 4.1. There are lots of other examples of 4-vectors, and we won't try to list them here.

If you take the four components of a 4-vector, you can make a scalar out of them. We already did this when we constructed the scalar $(d\tau)^2$ from the 4-vector dX^μ:

$$(d\tau)^2 = (dt)^2 - (dx)^2 - (dy)^2 - (dz)^2.$$

We can follow the same procedure with *any* 4-vector. If A^μ is a 4-vector, then the quantity

$$(A^0)^2 - (A^x)^2 - (A^y)^2 - (A^z)^2$$

is a scalar for exactly the same reasons. Once you know that the A^μ components transform the same way as t and x, you can see that the difference of the squares of the time component and the space component will not change under a Lorentz transformation. You can show this using the same algebra we used with $(d\tau)^2$.

We've seen how to construct a scalar from a 4-vector. Now we'll do the opposite, that is, construct a 4-vector from a scalar. We do this by differentiating the scalar with respect to each of the four components of space and time. Together, those four derivatives form a 4-vector. If we have a scalar ϕ, the quantities

$$\frac{\partial\phi}{\partial X^\mu} = \left(\frac{\partial\phi}{\partial X^0}, \frac{\partial\phi}{\partial X^1}, \frac{\partial\phi}{\partial X^2}, \frac{\partial\phi}{\partial X^3} \right)$$

are the components of a (covariant, or lower) 4-vector.

4.4.3 Building a Relativistic Lagrangian

We now have a set of tools for creating Lagrangians. We know how things transform, and we know how to construct scalars from vectors and other objects. How do we construct a Lagrangian? It's simple. The Lagrangian itself—the thing that we add up over all these little cells to form an action integral—must be the same in every coordinate frame. In other words, it must be a scalar! That's all there is to it. You take a field ϕ and consider all the possible scalars you can make from it. These scalars are candidate building blocks for our Lagrangian.

Let's look at some examples. Of course, ϕ itself is a scalar in this example, but so is any function of ϕ. If everyone agrees on the value of ϕ, they will also agree on the values of ϕ^2, $4\phi^3$, $\sinh(\phi)$, and so on. Any function of ϕ, for example a potential energy $V(\phi)$, is a scalar and therefore is a candidate for inclusion in a Lagrangian. In fact, we've already seen plenty of Lagrangians that incorporate $V(\phi)$.

What other ingredients could we use? Certainly we want to include derivatives of the field. Without them, our field theory would be trivial and uninteresting. We just need to be sure to put the derivatives together in a way that produces a scalar. But that's easy! First, we use the derivatives to build the 4-vector,

$$\frac{\partial \phi}{\partial X^\mu}.$$

Next, we use the components of our 4-vector to construct a scalar. The resulting scalar is

$$\left(\frac{\partial \phi}{\partial t}\right)^2 - \left(\frac{\partial \phi}{\partial x}\right)^2 - \left(\frac{\partial \phi}{\partial y}\right)^2 - \left(\frac{\partial \phi}{\partial z}\right)^2.$$

Here is a nontrivial expression we can put into a Lagrangian. What else could we use? Certainly, we can multiply by a numerical constant. For that matter, we can multiply by any function of any scalar. Multiplying two things that are invariant produces a third invariant. For example, the expression

$$\left[\left(\frac{\partial \phi}{\partial t}\right)^2 - \left(\frac{\partial \phi}{\partial x}\right)^2 - \left(\frac{\partial \phi}{\partial y}\right)^2 - \left(\frac{\partial \phi}{\partial z}\right)^2\right] F(\phi)$$

would be a legal Lagrangian. It's somewhat complicated, and we won't develop it here, but it qualifies as a Lorentz invariant Lagrangian. We could do something even uglier, like taking the expression inside the square brackets and raising it to some power. Then we'd have higher powers of derivatives. That would be ugly, but it would still be a legal Lagrangian.

What about higher-*order* derivatives? In principle, we could use them if we turned them into scalars. But that would take us outside the confines of classical mechanics. Within the scope of classical mechanics, we can use functions of the coordinates and their first derivatives. Higher *powers* of first derivatives are acceptable, but higher derivatives are not.

4.4.4 Using Our Lagrangian

Despite the restrictions of classical mechanics and the requirement of Lorentz invariance, we still have a tremendous amount of freedom in choosing a Lagrangian to work with. Let's have a close look at this one:

$$\mathcal{L} = \frac{1}{2}\left[\frac{1}{c^2}\left(\frac{\partial\phi}{\partial t}\right)^2 - \left(\frac{\partial\phi}{\partial x}\right)^2 - \left(\frac{\partial\phi}{\partial y}\right)^2 - \left(\frac{\partial\phi}{\partial z}\right)^2\right] - \frac{\mu^2}{2}\phi^2.$$
(4.24)

This is essentially the same as the Lagrangian of Eq. 4.7, and now you can see why I chose that one for our nonrelativistic example. The new factor $\frac{1}{c^2}$ in the first term just switches us back to conventional units. I also replaced the generic potential function $V(\phi)$ with a more explicit function, $-\frac{\mu^2}{2}\phi^2$. The factor $\frac{1}{2}$ is nothing more than a convention with no physical meaning. It's the same $\frac{1}{2}$ that appears in the expression $\frac{1}{2}mv^2$ for kinetic energy. We could have taken this to be mv^2 instead; if we did, then our mass would differ from Newton's mass by a factor of two.

Back in Lecture 1, we explained that a general Lorentz transformation is equivalent to a combination of space rotations, together with a simple Lorentz transformation along the x axis. We've shown that Eq. 4.24 is invariant under a simple Lorentz transformation. Is it also invariant with respect to space rotations? The answer is yes, because the spatial part of the expression is the sum of the squares of the components of a space vector. It behaves like the expression $x^2 + y^2 + z^2$ and is therefore rotationally invariant. Because the Lagrangian of Eq. 4.24 is invariant not only with respect to Lorentz transformations along the x axis but also with respect to rotations of space, it's invariant under a general Lorentz transformation.

Eq. 4.24 is one of the simplest field theories.[4] It gives rise to a wave equation in the same way as Eq. 4.7 does. Following the same pattern we did for that example, it's not hard to see that the wave equation derived from Eq. 4.24 is

$$\frac{1}{c^2}\frac{\partial^2\phi}{\partial t^2} - \frac{\partial^2\phi}{\partial y^2} - \frac{\partial^2\phi}{\partial x^2} - \frac{\partial^2\phi}{\partial z^2} + \mu^2\phi = 0. \qquad (4.25)$$

It's a particularly simple wave equation because it's linear; the field and its derivatives only appear to the first power. There are no terms like ϕ^2 or ϕ times a derivative of ϕ. We get an even simpler version when we set μ equal to zero:

$$\frac{1}{c^2}\frac{\partial^2\phi}{\partial t^2} - \frac{\partial^2\phi}{\partial y^2} - \frac{\partial^2\phi}{\partial x^2} - \frac{\partial^2\phi}{\partial z^2} = 0. \qquad (4.26)$$

4.4.5   Classical Field Summary

We now have a process for developing a classical field theory. Our first example was a scalar field, but the same process could apply to a vector or even a tensor field. Start with the field itself. Next, figure out all the scalars you can make, using the field itself and its derivatives. Once you have listed or characterized all the scalars you can create, build a Lagrangian from functions of those scalars, for example, sums of terms. Next, apply the Euler-Lagrange equations.

---

[4] We could make it even simpler by dropping the last term, $-\frac{\mu^2}{2}\phi^2$.

That amounts to writing the equations of motion, or the field equations describing the propagation of waves, or whatever else the field theory is supposed to describe. The next step is to study the resulting wave equation.

Classical fields need to be continuous. A field that's not continuous would have infinite derivatives, and therefore infinite action. Such a field would also have infinite energy. We'll have more to say about energy in the next lecture.

4.5 Fields and Particles—A Taste

Before wrapping up, I want to say a few things about the relation between particles and fields.[5] If I had worked out the rules for electrodynamics instead of a simple scalar field, I might tell you how charged particles interact with an electromagnetic field. We haven't done that yet. But how might a particle interact with a scalar field ϕ? How might the presence of the scalar field affect the motion of a particle?

Let's think about the Lagrangian of a particle moving in the presence of a preestablished field. Suppose someone has solved the equations of motion, and we know that the field $\phi(t, x)$ is some specific function of time and space. Now consider a particle moving in that field. The particle might be coupled to the field, in some way analogous to a charged particle in an electromagnetic field. How does the particle move? To answer that question, we go back to particle mechanics in the presence of a field. We've already written down a Lagrangian for a particle. It was $-m d\tau$, because $d\tau$ is just about the only invariant quantity available. The action

---

[5] Not the quantum mechanical relation between particles and fields, just the interaction between ordinary classical particles and classical fields.

integral was

$$Action = -m \int d\tau. \tag{4.27}$$

To get the correct nonrelativistic answers for low velocities, we found that we need the minus sign, and that the parameter m behaves like the nonrelativistic mass. Using the relationship $d\tau^2 = dt^2 - dx^2$, we rewrote this integral as

$$Action = -m \int \sqrt{dt^2 - dx^2},$$

where dx^2 stands for all three directions of space. Then, we factored out the dt^2, and our action integral became

$$Action = -m \int dt\sqrt{1 - \left(\frac{dx}{dt}\right)^2}.$$

Noticing that dx/dt is a synonym for velocity, this became

$$Action = -m \int dt\sqrt{1 - v^2}.$$

We then expanded the Lagrangian $-m\sqrt{1 - v^2}$ as a power series and found that it matches the familiar classical Lagrangian at low velocities. But this new Lagrangian is relativistic.

What can we do to this Lagrangian to allow the particle to couple to a field? For the field to affect the particle, the field itself needs to appear somewhere in the Lagrangian. We need to insert it in a way that is Lorentz invariant. In other words, we have to construct a scalar from the field. As we saw before, there are many ways to accomplish this, but one simple action we could try is

$$Action = -\int [m + \phi(t, x)]d\tau$$

or

$$Action = -\int [m + \phi(t, x)]\sqrt{1 - v^2}\, dt. \qquad (4.28)$$

This corresponds to a Lagrangian,

$$\mathcal{L} = -[m + \phi(t, x)]\sqrt{1 - v^2}. \qquad (4.29)$$

This is one of the simplest things you can do, but there are lots of other possibilities. For example, in the preceding Lagrangian, you could replace $\phi(t, x)$ with its square, or with any other function of $\phi(t, x)$. For now, we'll just use this simple Lagrangian, and its corresponding action integral, Eq. 4.28.

Eq. 4.29 is one possible Lagrangian for a particle moving in a preestablished field. Now we can ask: How does the particle move in this field? This is similar to asking how a particle moves in an electric or magnetic field. You write down a Lagrangian for the particle in the electric or magnetic field; you don't worry about how the field got there. Instead, you just write down the Lagrangian and then write out the Euler-Lagrange equations. We'll work out some of the details for this example. But before we do, I want to point out an interesting feature of Eq. 4.29.

4.5.1 The Mystery Field

Suppose, for some reason, the field $\phi(t, x)$ tends to migrate to some specific constant value other than zero. It just happens to "like" getting stuck at that particular value. In that case, $\phi(t, x)$ would be constant or approximately constant, despite its formal dependence on t and x. The motion of the particle would then look exactly the same as the motion of a particle whose mass is $m + \phi$. Let me say this again: The particle with mass m would behave as though its mass is $m + \phi$. This is the simplest example of a scalar field that gives rise to a shift in the mass of a particle.

At the beginning of the lecture, I mentioned that there's a field in nature that closely resembles the one we're looking at today, and I invited you to try guessing which one it is. Have you figured it out? The field we're looking at bears a close resemblance to the Higgs field. In our example, a shift in the value of a scalar field shifts the masses of particles. Our example is not exactly the Higgs mechanism, but it's closely related. If a particle starts out with a mass of zero and is coupled to the Higgs field, this coupling can effectively shift the particle's mass to a nonzero value. This shifting of mass values is roughly what people mean when they say that the Higgs field gives a particle its mass. The Higgs field enters into the equations *as if* it were part of the mass.

4.5.2   Some Nuts and Bolts

We'll wrap up this lecture with a quick peek at the Euler-Lagrange equations for our scalar field example. We're not going to follow them all the way through because they become ugly after the first few steps. We'll just give you a taste of this process for now. For simplicity, we'll work as if there's only one direction of space, with the particle moving only in the x direction. I'll copy our Lagrangian, Eq. 4.29, here for easy reference, replacing the variable v with $\dot{x}$:

$$\mathcal{L} = -[m + \phi(t, x)]\sqrt{1 - \dot{x}^2}.$$

The first step in applying Lagrange's equations is to calculate the partial derivative of $\mathcal{L}$ with respect to $\dot{x}$. Remember that when we take a partial derivative with respect to some variable, we temporarily regard all the other variables as constants. In this case, we regard the expression in square brackets as a constant because it has no explicit dependence on $\dot{x}$. On the other hand, the expression $\sqrt{1 - \dot{x}^2}$ does depend explicitly on $\dot{x}$. Taking the partial derivative of this expression results in

$$\frac{\partial \mathcal{L}}{\partial \dot{x}} = \frac{[m + \phi(t, x)]\dot{x}}{\sqrt{1 - \dot{x}^2}}. \qquad (4.30)$$

This equation should look familiar. We obtained a nearly identical result in Lecture 3 when we did a similar calculation to find the momentum of a relativistic particle. The only thing that's different here is the extra term $\phi(t, x)$ inside the square brackets. This supports the notion that the expression $[m + \phi(t, x)]$ behaves like a *position-dependent* mass.

Continuing with the Euler-Lagrange equations, the next step is to differentiate Eq. 4.30 with respect to time. We'll just indicate this operation symbolically by writing

$$\frac{d}{dt} \frac{\partial \mathcal{L}}{\partial \dot{x}} = \frac{d}{dt} \frac{[m + \phi(t, x)]\dot{x}}{\sqrt{1 - \dot{x}^2}}.$$

That's the left side of the Euler-Lagrange equation. Let's look at the right side, which is

$$\frac{\partial \mathcal{L}}{\partial x}.$$

Since we're differentiating with respect to x, we ask whether $\mathcal{L}$ depends explicitly on x. It does, because $\phi(t, x)$ depends on x, and ϕ happens to be the *only* place where x appears. The resulting partial derivative is

$$\frac{\partial \mathcal{L}}{\partial x} = -\frac{\partial \phi}{\partial x} \sqrt{1 - \dot{x}^2},$$

and therefore the Euler-Lagrange equation becomes

$$\frac{d}{dt} \frac{[m + \phi(t,x)]\dot{x}}{\sqrt{1 - \dot{x}^2}} = -\frac{\partial \phi}{\partial x} \sqrt{1 - \dot{x}^2}.$$

That's the equation of motion. It's a differential equation that describes the motion of the field. If you try to work out the time derivative on the left side, you'll see that it's quite complicated. We'll stop at this point, but you may want to think about how to incorporate the velocity of light c into this equation, and how the field behaves in the nonrelativistic limit of low velocities. We'll have some things to say about that in the next lecture.

Lecture 5

Particles and Fields

Nov 9, 2016.

The day after election day. Art is morosely staring into his beer. Lenny is staring into his glass of milk. No one—not even Wolfgang Pauli—can think of anything funny to say. But then mighty John Wheeler rises to his feet and raises his hand at the bar where all can see him. Sudden silence and John speaks:

"Ladies and gentlemen, in this time of terrible uncertainty, I want to remind you of one sure thing: SPACETIME TELLS MATTER HOW TO MOVE; MATTER TELLS SPACETIME HOW TO CURVE."

"Bravo!"

A cheer is heard in Hermann's Hideaway and things brighten. Pauli raises his glass:

"To Wheeler! I think he's hit the nail on the head. Let me try to say it even more generally: FIELDS TELL CHARGES HOW TO MOVE; CHARGES TELL FIELDS HOW TO VARY."

In quantum mechanics, fields and particles are the same thing. But our main topic is *classical* field theory, where fields are fields, and particles are particles. I won't say "never the twain will meet" because they will meet; in fact, they'll meet very soon. But they're not the same thing. The central question of this lecture is this:

If a field affects a particle, for example by creating forces on it, must the particle affect the field?

If A affects B, why must B necessarily affect A? We'll see that the two-way nature of interactions, often called "action and reaction," is built into the Lagrangian action principle. As a simple example, suppose we have two coordinates, x and y, along with an action principle.[1] Generally, the Lagrangian will depend on x and y, and also on $\dot{x}$ and $\dot{y}$. One possibility is that the Lagrangian is simply a sum of two terms: a Lagrangian for x and $\dot{x}$, plus a separate Lagrangian for y and $\dot{y}$

$$\mathcal{L} = \mathcal{L}_x(x, \dot{x}) + \mathcal{L}_y(y, \dot{y}), \qquad (5.1)$$

where I've labeled the Lagrangians on the right side with subscripts to show that they may be different. Let's look at the Euler-Lagrange equation (which is the equation of motion) for x,

$$\frac{d}{dt} \frac{\partial \mathcal{L}}{\partial \dot{x}} - \frac{\partial \mathcal{L}}{\partial x} = 0.$$

Because $\mathcal{L}_y(y, \dot{y})$ has no dependency on x or on $\dot{x}$, the $\mathcal{L}_y$ terms drop out, and the Euler-Lagrange equation for the x

---

[1] Of course we could have many more than two, and they don't need to be orthogonal Cartesian coordinates.

coordinate becomes

$$\frac{d}{dt}\frac{\partial \mathcal{L}_x}{\partial \dot{x}} - \frac{\partial \mathcal{L}_x}{\partial x} = 0.$$

The y variable and its time derivative do not appear at all. Likewise, the equation of motion for the y variable will not include any reference to x or $\dot{x}$. As a result, x does not affect y, and y does not affect x. Let's look at another example,

$$\mathcal{L} = \frac{1}{2}(\dot{x})^2 + \frac{1}{2}(\dot{y})^2 - V_x(x) - V_y(y),$$

where the V_x and V_y terms are potential energy functions. Once again, the x and y coordinates of this Lagrangian are completely separated. The Lagrangian is a sum of terms that only involve x or only involve y, and by a similar argument the x and y coordinates will not affect each other.

But suppose we know that y *does* affect x. What would that tell us about the Lagrangian? It tells us that the Lagrangian must be more complicated; that there must be things in it that somehow affect both x and y. To write such a Lagrangian, we need to put in some additional ingredient that involves both x and y in a way that you can't unravel. For example, we could just add an xy term:

$$\mathcal{L} = \frac{1}{2}(\dot{x})^2 + \frac{1}{2}(\dot{y})^2 - V_x(x) - V_y(y) + xy.$$

This guarantees that the equation of motion for x will involve y, and vice versa. If x and y appear in the Lagrangian coupled together in this manner, there's no way for one to affect the other without the other affecting the one. It's as simple as that. That's the reason A must affect B if B affects A.

In the previous lecture we looked at a simple field and asked how it affects a particle. After a brief review of that example, we'll ask the opposite question: How does the particle affect the field? This is very much an analog of electromagnetic interactions, where electric and magnetic fields affect the motion of charged particles, and charged particles create and modify the electromagnetic field. The mere presence of a charged particle creates a Coulomb field. These two-way interactions are not two independent things; they come from the same Lagrangian.

5.1 Field Affects Particle (Review)

Let's start out with a given field,

$$\phi(t, x),$$

that depends on t and x. For now, assume that ϕ is some known function. It may or may not be a wave. We're not going to ask about the dynamics of ϕ just yet; first we'll look at the Lagrangian for the particle. Recall from previous lectures (Eq. 3.24, for example) that this Lagrangian is

$$\mathcal{L}_{particle} = -m\sqrt{1 - \dot{x}^2}.$$

I've labeled it $\mathcal{L}_{particle}$ for clarity. Apart from the factor $-m$, the action $\mathcal{L}_{particle}$ amounts to a sum of all the proper times along each little segment of a path. In the limit, as the segments become smaller and smaller, the sum becomes an integral

$$\int \mathcal{L}_{particle}\, dt = \int -m\sqrt{1 - \dot{x}^2}\, dt$$

over the coordinate time t. This is the same as $-m$ times the integral of the proper time $d\tau$ from one end of the trajectory to the other. This Lagrangian doesn't contain anything that causes the field to affect the particle. Let's modify it in a simple way, by adding the field value $\phi(t, x)$ to m.

$$\int \mathcal{L}_{particle}\, dt = \int -[m + \phi(t, x)]\sqrt{1 - \dot{x}^2}\, dt.$$

We can now work out the Euler-Lagrange equations and find out how $\phi(t, x)$ affects the motion of the particle. Rather than work out the full relativistic equations of motion, we'll look at the nonrelativistic limit where the particle moves very slowly. That's the limit in which the speed of light goes to infinity—when c is far bigger than any other velocity in the problem. It's helpful to restore the constant c in our equations to see how this works. The modified action integral is

$$\int \mathcal{L}_{particle}\, dt = \int -[mc^2 + g\phi(t, x)]\sqrt{1 - \frac{\dot{x}^2}{c^2}}\, dt.$$

We can check this for dimensional consistency. The Lagrangian has units of energy, and so does $-mc^2$. I've multiplied $\phi(t, x)$ by a constant g, which is called a *coupling constant*. It measures the strength by which the field affects the motion of the particle, and we can select its units to guarantee that g times $\phi(t, x)$ has units of energy. So far, g can be anything; we don't know its value. Both terms inside the square root are pure numbers.

Now let's expand the square root using the approximation formula

$$(1 - \epsilon)^{\frac{1}{2}} \approx 1 - \frac{\epsilon}{2},$$

where ϵ is a small number. Re-writing the square root with an exponent of $\frac{1}{2}$, and equating $\frac{\dot{x}^2}{c^2}$ with ϵ, we can see that

$$\sqrt{1 - \frac{\dot{x}^2}{c^2}} = \left(1 - \frac{\dot{x}^2}{c^2}\right)^{\frac{1}{2}} \approx 1 - \frac{\dot{x}^2}{2c^2}.$$

The higher-order terms are far smaller because they involve higher powers of the ratio $\dot{x}^2/c^2$. We can now use this approximate expression to replace the square root in the action integral, resulting in

$$\int \mathcal{L}_{particle}\, dt = \int -[mc^2 + g\phi(t,x)]\left(1 - \frac{\dot{x}^2}{2c^2}\right) dt. \quad (5.2)$$

Let's look at this integral and find the biggest terms—the terms that are most important when the speed of light gets large. The first term, mc^2, is just a number. When you take derivatives of the Lagrangian it just "comes along for the ride" and has no meaningful impact on the equations of motion. We'll ignore it. In the next term,

$$(mc^2)\left(\frac{\dot{x}^2}{2c^2}\right) = \frac{m\dot{x}^2}{2},$$

the speed of light cancels itself out altogether. Therefore this term is part of the limit in which the speed of light goes to infinity, and it survives that limit. This term is quite familiar; it's our old friend, the nonrelativistic kinetic energy. The term

$$g\phi(t,x)\left(\frac{\dot{x}^2}{2c^2}\right)$$

becomes zero in the limit of large c because of the c^2 in its denominator; we ignore it. Finally, the term $g\phi(t,x)$ contains no speeds of light. Therefore it survives, and we keep it in the Lagrangian, which now becomes

$$\mathcal{L}_{particle} = \frac{m\dot{x}^2}{2} - g\phi(t,x). \quad (5.3)$$

That's all there is when the particle moves slowly. We can now compare it with the old fashioned nonrelativistic

Lagrangian, kinetic energy minus potential energy,

$$T - V.$$

We've already recognized the first term of Eq. 5.3 as kinetic energy, and now we can identify $g\phi(t, x)$ as the potential energy of a particle in this field. The constant g indicates the strength of coupling between the particle and the field. In electromagnetism, for example, the strength of coupling of the field to a particle is just the electric charge. The bigger the electric charge, the bigger the force on a particle in a given field. We'll come back to this idea.

5.2   Particle Affects Field

How does the particle affect the field? The important thing to understand is that there's only *one* action, the "total" action. The total action includes action for the field *and* action for the particle. I can't emphasize this enough: What we're studying is a combined system that consists of a) a field and b) a particle moving through the field.

Fig. 5.1 illustrates the physics problem we're trying to solve. It shows a region of spacetime, represented as a cube.[2] Time points upward and the x axis points to the right. Inside this region, there's a particle that travels from one spacetime point to another. The two dots are the end points of its trajectory. We also have a field $\phi(t, x)$ inside the region. I wish I could think of a clever way to draw this field without cluttering the diagram, because it's every bit as physical as the particle; it's part of the system.[3]

---

[2] With too few dimensions, of course.

[3] In the video, Leonard shows the field by using a colored marker to fill the region with "red mush." Our diagrams don't use color, so we have to settle for "imaginary red mush." Just think of your least favorite school cafeteria entrée. –AF

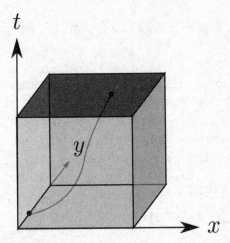

Figure 5.1: Particle moving through a region of spacetime filled with "red mush"—the scalar field $\phi(t, x)$.

To find out how this field-particle system behaves, we need to know the Lagrangian and minimize the action. In principle, this is simple; we just vary the parameters of the problem until we find a set of parameters that results in the smallest possible action. We wiggle the field around in different ways, and we wiggle the particle trajectory between its two end points until the action integral is as small as we can make it. That gives us the trajectory *and* the field that satisfy the principle of least action.

Let's write down the whole action, the action that includes both the field and the particle. First, we need an action for the field,

$$Action_{field} = \int \mathcal{L}_{field} d^4 x,$$

where the symbol $\mathcal{L}_{field}$ means "Lagrangian for the field."

This action integral is taken over the entire spacetime region, t, x, y, and z. The symbol d^4x is shorthand for $dt\,dx\,dy\,dz$. For this example, we'll base our field Lagrangian on Eq. 4.7, a Lagrangian that we used in Lecture 4. We'll use a simplified version that only references the x direction in space and has no potential function $V(\phi)$.[4] The Lagrangian

$$\mathcal{L}_{field} = \frac{1}{2}\left(\frac{\partial \phi}{\partial t}\right)^2 - \frac{1}{2}\left(\frac{\partial \phi}{\partial x}\right)^2$$

leads to the action integral

$$Action_{field} = \int \frac{1}{2}\left(\frac{\partial \phi}{\partial t}\right)^2 - \frac{1}{2}\left(\frac{\partial \phi}{\partial x}\right)^2 dx. \qquad (5.4)$$

This is the action for the field.[5] It doesn't involve the particle at all. Now let's incorporate the action for the particle, Eq. 5.2 with the speeds of light removed. This is just

$$Action_{particle} = \int \mathcal{L}_{particle}\, dt,$$

or

$$Action_{particle} = \int -[m + g\phi(t, x)]\left(1 - \frac{\dot{x}^2}{2}\right) dt. \qquad (5.5)$$

Although $Action_{particle}$ is the particle action, it also depends on the field. This is important; when we wiggle the field, this action varies. In fact, it is wrong to think of $Action_{particle}$ purely as a particle action: The term

$$g\phi(t, x)\left(1 - \frac{\dot{x}^2}{2}\right)$$

---

[4] Or, equivalently, $V(\phi) = 0$.

[5] To avoid clutter, I'm using only one space coordinate.

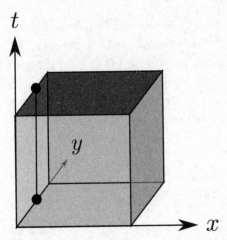

Figure 5.2: Particle at Rest in Imaginary Red Mush (black ink on paper).

is an interaction term that tells the particle how to move in the field, but it also tells the field how to vary in the presence of the particle. From now on I will call this term $\mathcal{L}_{interaction}$ and cease thinking of it as having to do only with the particle.

We'll consider the simple special case where the particle is at rest at $x = 0$. It's reasonable to assume that there's some solution where the particle is at rest—classical particles do rest sometimes. Fig. 5.2 shows the spacetime trajectory of a resting particle. It's just a vertical line.

How do we modify Eq. 5.5 to show that the particle is at rest? We just set $\dot{x}$, the velocity, equal to zero. The simplified action integral is

$$\int \mathcal{L}_{interaction}\, dt = \int [-g\phi(t, x)]\, dt.$$

Because the particle sits at the fixed position $x = 0$, we can replace $\phi(t, x)$ with $\phi(t, 0)$ and write

$$\int \mathcal{L}_{interaction} \, dt = \int -g\phi(t, 0) \, dt$$

$$Action_{interaction} = \int -g\phi(t, 0) \, dt. \qquad (5.6)$$

Let's take a closer look at $\mathcal{L}_{interaction}$. Notice that it only depends on the value of the field at the origin. More generally it depends on the value of the field at the location of the particle. Nevertheless when the field is wiggled, $\phi(t, 0)$ will wiggle, affecting the action. As we will see, this affects the equation of motion for the field.

The action for a field is normally written as an integral over space and time, but Eq. 5.6 is an integral that only runs over time. There is nothing wrong with it, but it's convenient to rewrite it as an integral over space and time. We'll use a trick that involves the idea of a *source function*. Let $\rho(x)$ be a fixed definite function of space, but for the moment not time. (I'll give you a hint: $\rho(x)$ is something like a charge density.) Let's forget the particle but replace it with a term in the Lagrangian that I'll continue to call $\mathcal{L}_{interaction}$,

$$\mathcal{L}_{interaction} = -g\rho(x)\phi(t, x). \qquad (5.7)$$

The corresponding term in the action for the field is

$$Action_{interaction} = -\int d^4x \, g\rho(x)\phi(t, x). \qquad (5.8)$$

This looks quite different from the action in Eq. 5.2. To make them the same, we use the trick that Dirac invented—the Dirac delta function $\delta(x)$. The delta function is a function of the space coordinates that has a peculiar property. It is zero everywhere but at $x = 0$. Nevertheless it has a nonzero integral,

$$\int \delta(x)\,dx = 1. \tag{5.9}$$

Let's imagine graphing the delta function. It is zero everywhere except in the immediate vicinity of $x = 0$. But it is so large in that vicinity that it has a total area equal to 1. It is a very high and narrow function, so narrow that we may think of it as concentrated at the origin, but so high that it has a finite area.

No real function behaves that way, but the Dirac function is not an ordinary function. It is really a mathematical rule. Whenever it appears in an integral multiplying another function $F(x)$, it picks out the value of F at the origin. Here is its mathematical definition:

$$\int F(x)\delta(x)\,dx = F(0), \tag{5.10}$$

where $F(x)$ is an "arbitrary" function.[6] If you have some function $F(x)$ and integrate it as shown, the delta function picks out the value of $F(x)$ at $x = 0$. It filters out all other values. It's the analog of the Kronecker delta, but operates on continuous functions. You can visualize $\delta(x)$ as a function whose value is zero everywhere, except when x gets very close to zero. When x is close to zero, $\delta(x)$ has a very tall spike. For our current problem we need a three-dimensional delta function that we call $\delta^3(x, y, z)$. We define $\delta^3(x, y, z)$ as the product of three one-dimensional delta functions,

$$\delta^3(x, y, z) = \delta(x)\delta(y)\delta(z).$$

---

[6] $F(x)$ is not strictly arbitrary. However, it represents a broad class of functions, and we can think of it as arbitrary for our purposes.

As we do elsewhere, we often use the shorthand $\delta^3(x)$, where x represents all three directions of space. Where is $\delta^3(x, y, z)$ nonzero? It's nonzero where all three factors on the right side are nonzero, and that only happens in one place: the point $x = 0, y = 0, z = 0$. At that point of space it's enormous.

The trick in writing the interaction term as an integral should now be fairly obvious. We simply replace the particle with a source function ρ that we choose to be a delta function,

$$\rho(x) = \delta^3(x). \qquad (5.11)$$

The action in Eq. 5.8 then takes the form

$$Action_{interaction} = \int -g\phi(t, x)\delta^3(x)\, d^4x. \qquad (5.12)$$

The point is that if we integrate this over the space coordinates, the delta function rule tells us to simply evaluate $g\phi(x)$ at the origin, and we get something we've seen before:

$$Action_{interaction} = -g \int \phi(t, 0)\, dt. \qquad (5.13)$$

Now let's combine the field action with the interaction term. We do that in the simplest possible way, by just adding the two of them together. Combining the action terms of Eqs. 5.4 and 5.12 results in

$$Action_{total} = \int \left[\frac{1}{2}\left(\frac{\partial\phi}{\partial t}\right)^2 - \frac{1}{2}\left(\frac{\partial\phi}{\partial x}\right)^2 - g\rho(x)\phi(t, x) \right] d^4x,$$

or replacing the source function with the delta function,

$$Action_{total} = \int \left[\frac{1}{2}\left(\frac{\partial\phi}{\partial t}\right)^2 - \frac{1}{2}\left(\frac{\partial\phi}{\partial x}\right)^2 - g\delta^3(x)\phi(t, x) \right] d^4x.$$

Here it is, the total action as an integral over space and time.

The expression inside the big square brackets is the Lagrangian. Like any good field Lagrangian, it has the field action (represented here by partial derivatives) along with a delta function term that represents the effect of the particle on the field. This particular delta function represents the special situation where the particle is at rest. But we could also jazz it up so that the particle moves. We could do that by making the delta function move around with time. But that's not important right now. Instead, let's work out the equations of motion, based on the Lagrangian

$$\mathcal{L}_{total} = \left[\frac{1}{2} \left(\frac{\partial \phi}{\partial t} \right)^2 - \frac{1}{2} \left(\frac{\partial \phi}{\partial x} \right)^2 - g\rho(x)\phi(t,x) \right]. \quad (5.14)$$

5.2.1 Equations of Motion

For convenience, I'll rewrite Eq. 4.5, the Euler-Lagrange equation, right here.

$$\sum_{\mu} \frac{\partial}{\partial X^\mu} \frac{\partial \mathcal{L}}{\partial \left(\dfrac{\partial \phi}{\partial X^\mu} \right)} - \frac{\partial \mathcal{L}}{\partial \phi} = 0. \quad (5.15)$$

Eq. 5.15 tells us what to do with $\mathcal{L}_{total}$ in order to find the equations of motion. The index μ runs through the values 0, 1, 2, and 3. The first value it takes is 0, which is the time component. Therefore our first step is to find the derivative

$$\frac{\partial}{\partial t} \frac{\partial \mathcal{L}_{total}}{\partial \left(\dfrac{\partial \phi}{\partial t} \right)}.$$

Let's unwind this calculation step by step. First, what is the partial of $\mathcal{L}_{total}$ with respect to $\frac{\partial \phi}{\partial t}$? There's only one term in $\mathcal{L}_{total}$ that involves $\frac{\partial \phi}{\partial t}$. Straightforward differentiation shows that the result is

$$\frac{\partial \mathcal{L}_{total}}{\partial \left(\dfrac{\partial \phi}{\partial t} \right)} = \frac{\partial \phi}{\partial t}.$$

Applying $\dfrac{\partial}{\partial t}$ to each side results in

$$\frac{\partial}{\partial t} \frac{\partial \mathcal{L}}{\partial \left(\dfrac{\partial \phi}{\partial t} \right)} = \frac{\partial^2 \phi}{\partial t^2}.$$

That's the first term in the equation of motion. It looks a lot like an acceleration. We started with a kinetic energy term that amounts to $\dot{\phi}^2$. When we worked out the derivative, we got something that looks like an acceleration of ϕ.

Next, we let μ take the value 1 and calculate the first space component. The form of these terms is exactly the same as the form of the time component except for the minus sign. So the first two terms of the equation of motion are

$$\frac{\partial^2 \phi}{\partial t^2} - \frac{\partial^2 \phi}{\partial x^2}.$$

Because the y and z components have the same form as the x component, we can add them in as well,

$$\frac{\partial^2 \phi}{\partial t^2} - \frac{\partial^2 \phi}{\partial x^2} - \frac{\partial^2 \phi}{\partial y^2} - \frac{\partial^2 \phi}{\partial z^2}.$$

If those were the only terms, I would set this expression equal to zero, and we would have a good old-fashioned wave equation for ϕ. However, the Lagrangian depends on ϕ in another way because of the interaction term. The complete equation of motion is

$$\frac{\partial^2 \phi}{\partial t^2} - \frac{\partial^2 \phi}{\partial x^2} - \frac{\partial^2 \phi}{\partial y^2} - \frac{\partial^2 \phi}{\partial z^2} = \frac{\partial \mathcal{L}_{total}}{\partial \phi},$$

and we can see that the last term is

$$\frac{\partial \mathcal{L}_{total}}{\partial \phi} = -g\rho(x).$$

Adding this final piece to the equation of motion gives us

$$\frac{\partial^2 \phi}{\partial t^2} - \frac{\partial^2 \phi}{\partial x^2} - \frac{\partial^2 \phi}{\partial y^2} - \frac{\partial^2 \phi}{\partial z^2} = -g\rho(x). \tag{5.16}$$

Thus we see that the source function appears in the field equation for ϕ as an addition to the wave equation. Without the source term, $\phi = 0$ is a perfectly good solution, but that's not true when the source term is present. The source is literally that: a source of field that prevents the trivial $\phi = 0$ from being a solution.

In the actual case where the source is a particle at rest, $\rho(x)$ can be replaced by the delta function,

$$\frac{\partial^2 \phi}{\partial t^2} - \frac{\partial^2 \phi}{\partial x^2} - \frac{\partial^2 \phi}{\partial y^2} - \frac{\partial^2 \phi}{\partial z^2} = -g\delta^3(x). \tag{5.17}$$

Let's suppose for a moment that we're looking for static solutions to Eq. 5.16. After all, the particle is standing still, and there may be a solution where the field itself doesn't change with time. It seems plausible that a standing-still particle can create a field that also stands still—a field that does not vary with time. We might then look for a solution in which ϕ is time-independent. That means the $\dfrac{\partial^2 \phi}{\partial t^2}$ term in Eq. 5.16 would become zero. We can then change the signs of the remaining terms to plus and write

$$\frac{\partial^2 \phi}{\partial x^2} + \frac{\partial^2 \phi}{\partial y^2} + \frac{\partial^2 \phi}{\partial z^2} = +g\delta^3(x).$$

Perhaps you recognize this as Poisson's equation. It describes, among other things, the electrostatic potential of a point particle. It's often written by setting $\nabla^2\phi$ equal to a source (a charge density) on the right side.[7] In our example, the charge density is just a delta function. In other words, it's a high, sharp spike,

$$\nabla^2\phi = g\delta^3(x). \tag{5.18}$$

Of course, our current example is not an electric or magnetic field. Electrodynamics involves vector fields, and we're looking at a scalar field. But the similarities are striking.

The third term of the Lagrangian (Eq. 5.14) ties everything together. It tells the particle to move as if there were a potential energy $-g\phi(t, x)$. In other words, it exerts a force on the particle. This same term, when used for the equation of motion of the ϕ field, tells us that the ϕ field has a source.

These are not independent things. The fact that the field affects the particle tells us that the particle affects the field. For a particle at rest in a static field, Eq. 5.18 tells us exactly how. The parameter g determines how strongly the particle affects the field. The same parameter also tells us how strongly the field affects the particle. It makes a nice little story: Fields and particles affect each other through a common term in the Lagrangian.

5.2.2 Time Dependence

What happens if we allow the particle to move and the field to change with time? We'll confine ourselves to a single

---

[7] The symbol $\nabla^2\phi$ is shorthand for $\dfrac{\partial^2\phi}{\partial x^2} + \dfrac{\partial^2\phi}{\partial y^2} + \dfrac{\partial^2\phi}{\partial z^2}$. See Appendix B for a quick summary of the meaning of $\vec{\nabla}$ and other vector notation. You could also refer to Lecture 11 in *Volume I* of the Theoretical Minimum series. Many other references are available as well.

dimension of space. Eq. 5.18 becomes

$$\nabla^2\phi = g\delta(x).$$

In one dimension, the left side of this equation is just the second derivative of ϕ. Suppose I wanted to put the particle someplace else. Instead of putting it at the origin, suppose I want to put it at $x = a$? All I need to do is change x in the preceding equation to $x - a$, and the equation would become

$$\nabla^2\phi = g\delta(x - a).$$

The delta function $\delta(x - a)$ has its spike where x is equal to a. Suppose further that the particle is moving, and its position is a function of time, $a(t)$. We can write this as

$$\nabla^2\phi = g\delta(x - a(t)).$$

This would tell you that the field has a source, and that the source is moving. At any given time the source is at position $a(t)$. In this way we can accommodate a moving particle. But we still have one little wrinkle to deal with. If the particle is moving, we would *not* expect the field to be time independent. If the particle moves around, then the field must also depend on time. Remember the $\dfrac{\partial^2\phi}{\partial t^2}$ term that we zeroed out in the equation of motion (Eq. 5.16)? For a time-dependent field, we have to restore that term, resulting in

$$-\frac{\partial^2\phi}{\partial t^2} + \nabla^2\phi = g\delta(x - a(t)).$$

If the right side depends on time, there's no way to find a solution where ϕ itself is time-independent. The only way to make it consistent is to restore the term that involves time.

A moving particle, for example a particle that accelerates or vibrates, will give the field a time dependence. You probably know what it will do: It will radiate waves. But at the moment, we're not going to solve wave equations. Instead, we'll spend a little time talking about the notation of relativity.

5.3 Upstairs and Downstairs Indices

Notation is far more important than most people realize. It becomes part of our language and shapes the way we think— for better or for worse. If you're skeptical about this, just try switching to Roman numerals the next time you do your taxes.

The mathematical notation we introduce here makes our equations look simple and pretty. It saves us from having to write things like

$$\frac{\partial^2 \phi}{\partial t^2} - \frac{\partial^2 \phi}{\partial x^2} - \frac{\partial^2 \phi}{\partial y^2} - \frac{\partial^2 \phi}{\partial z^2}$$

every single time we write an equation.[8] In the previous lecture, we spent some time on standard notation for relativistic vectors, 4-vectors, and scalars. We'll revisit that material briefly, then explain the new condensed notation.

The symbol X^μ stands for the four coordinates of space-time, which we can write as

$$X^\mu = (t, x, y, z),$$

where the index μ runs over the four values 0 through 3. I'm going to start paying attention to where I put this index. Here, I've put the index up on top. That carries no meaning for now, but soon it will.

The quantity X^μ, if thought of as a displacement from the

---

[8] Substitute your favorite expletive for the word *single*.

origin, is a 4-vector. Calling it a 4-vector is a statement about the way it behaves under Lorentz transformation; any complex of four quantities that transforms the same way as t and x is a 4-vector.

The three space components of a 4-vector may equal zero in your reference frame. You, in your frame, would say that this displacement is purely timelike. But this is not an invariant statement. In my frame, the space components would not all equal zero, and I would say that the object does move in space. However, if *all four* components of a displacement 4-vector are zero in your frame, they will also be zero in my frame and in every other frame. A statement that all four components of a 4-vector are zero is an invariant statement.

Differences of 4-vectors are also 4-vectors. So are differential displacements such as

$$dX^\mu.$$

Starting with a 4-vector, we can make a scalar, a quantity that remains the same in every frame. For example, we can construct a proper time $d\tau^2$ from a displacement. You have already seen how. The quantity

$$(d\tau)^2 = dt^2 - dx^2 - dy^2 - dz^2$$

and its counterpart

$$(ds)^2 = -dt^2 + dx^2 + dy^2 + dz^2$$

are the same in every reference frame. They're scalars and have only one component. If a scalar is equal to zero in one frame of reference, it's zero in every frame. Indeed, this is the definition of a scalar; it's a thing that's the same in every frame.

The pattern we followed for combining the components of dX^μ to form the spacetime interval ds^2 (and the proper time

$d\tau$) is very general. We can apply it to *any* 4-vector A^μ, and the result will always be a scalar. We'll use this pattern over and over again, and we don't want to write the long expression

$$-(A^t)^2 + (A^x)^2 + (A^y)^2 + (A^z)^2$$

every time we need it. Instead, we'll create some new notation to make things easier. A matrix called the *metric* figures heavily in this new notation. In special relativity, the metric is often called η (Greek letter *eta*), with indices μ and ν. It's a simple matrix. In fact, it's almost the same as the identity matrix. It has three diagonal elements that are equal to 1, just like the unit matrix. But the fourth diagonal element is -1. This element corresponds to time. Here's what the entire matrix looks like:

$$\eta_{\mu\nu} = \begin{pmatrix} -1 & 0 & 0 & 0 \\ 0 & 1 & 0 & 0 \\ 0 & 0 & 1 & 0 \\ 0 & 0 & 0 & 1 \end{pmatrix}.$$

In this notation, we represent a 4-vector as a column. For example, the 4-vector A^ν is written as

$$A^\nu = \begin{pmatrix} A^t \\ A^x \\ A^y \\ A^z \end{pmatrix}.$$

Let's take the matrix $\eta_{\mu\nu}$ and multiply it by the vector A^ν, where ν runs from 0 to 3. Using the summation sign, we can write that as

$$\sum_\nu \eta_{\mu\nu} A^\nu = \eta_{\mu 0} A^0 + \eta_{\mu 1} A^1 + \eta_{\mu 2} A^2 + \eta_{\mu 3} A^3.$$

This is a new kind of object, which we'll call A_μ, with a *lower* index,

$$A_\mu = \sum_\nu \eta_{\mu\nu} A^\nu.$$

Let's figure out what this new object "A-with-a-lower-index" is. It's not the original 4-vector A^ν. If η really were the identity matrix, multiplying a vector by it would give you the exact same vector. But it's not the identity matrix. The -1 in the diagonal means that when you form the product $\sum_\nu \eta_{\mu\nu} A^\nu$, the sign of the first component, A^t, gets flipped, and everything else stays the same. We can write immediately that A_ν is

$$A_\nu = \begin{pmatrix} -A^t \\ A^x \\ A^y \\ A^z \end{pmatrix}.$$

In other words, when I perform this operation, I simply change the sign of the time component. In general relativity, the metric has a deeper geometric meaning. But for our purposes, it's just a convenient notation.[9]

5.4   Einstein Sum Convention

If necessity is the mother of invention, laziness is the father. The Einstein summation convention is an offspring of this happy marriage. We introduced it in Section 4.4.2, and now we explore its use a little further.

Whenever you see the same index both downstairs *and* upstairs in a single term, you automatically sum over that

---

[9] It might be better to express A_ν as a row matrix. However, the summation convention described in the next section minimizes the need to write out matrices in component form.

index. Summation is implied, and you don't need a summation symbol. For example, the term

$$A^\mu A_\mu \qquad\qquad (5.19)$$

means

$$A^0 A_0 + A^1 A_1 + A^2 A_2 + A^3 A_3$$

because the same index μ appears both upstairs and downstairs in the same term. On the other hand, the term

$$A_\nu A^\mu$$

does not imply summation, because the upstairs and downstairs indices are not the same. Likewise,

$$A_\nu A_\nu$$

does not imply summation even though the index ν is repeated, because both indices are downstairs.

You may recall that some of the equations in Section 3.4.3 used the symbol $\left(\dot{X}^i\right)^2$ to signify the sum of squares of space components. By using upstairs and downstairs indices along with the summation convention, we could have written

$$\dot{X}^i \dot{X}_i,$$

which is more elegant and precise.

The operation of Expression 5.19 has the effect of changing the sign of the time component. I should warn you that some authors follow the convention $(+1, -1, -1, -1)$ for the placement of these minus signs. I prefer the convention $(-1, 1, 1, 1)$, typically used by those who study general relativity.

An index that triggers the summation convention, like ν in the following example, doesn't have a specific value. It's called a *summation index* or a *dummy index*; it's a thing you sum over. By contrast, an index that is *not* summed over is called a *free index*. The expression

$$A_\mu = \eta_{\mu\nu} A^\nu \qquad (5.20)$$

depends on μ (which is a free index), but it doesn't depend on the summation index ν. If we replace ν with any other Greek letter, the expression would have exactly the same meaning. I should also mention that the terms *upstairs index* and *downstairs index* have formal names. An upper index is called *contravariant*, and a lower index is called *covariant*. I often use the simpler words *upper* and *lower*, but you should learn the formal terms as well. We can have A with an upper (contravariant) index, or A with a lower (covariant) index, and we use the matrix η to convert one to the other. Converting one kind of index to the other kind is called *raising* the index or *lowering* the index, depending on which way we go.

Exercise 5.1: Show that $A^\nu A_\nu$ has the same meaning as $A^\mu A_\mu$.

Exercise 5.2: Write an expression that undoes the effect of Eq. 5.20. In other words, how do we "go backwards"?

Let's have another look at the Expression 5.19,

$$A^\mu A_\mu.$$

This expression is summed over because it contains a repeated index, one upper and one lower. Previously, we expanded it using the indices 0 through 3. We can write the same expression using the labels t, x, y, and z:

$$A^\mu A_\mu = A^t A_t + A^x A_x + A^y A_y + A^z A_z.$$

For the three space components, the covariant and contra-variant versions are exactly the same. The first space component is just $(A^x)^2$, and it doesn't matter whether you put the index upstairs or downstairs. The same is true for the y and z components. But the time component becomes $-(A^t)^2$,

$$A^\mu A_\mu = -(A^t)^2 + (A^x)^2 + (A^y)^2 + (A^z)^2.$$

The time component has a minus sign because the operation of lowering or raising that index changes its sign. The contravariant and covariant time components have opposite signs, and A^t times A_t is $-(A^t)^2$. On the other hand, the contravariant and covariant space components have the *same* signs.

The quantity $A^\mu A_\mu$ is exactly what we think of as a scalar. It's the difference of the square of the time component and the square of the space component. If A^μ happens to be a displacement such as X^μ, then it's the same as the quantity τ^2, except with an overall minus sign; in other words, it's $-\tau^2$. But whatever sign it has, this sum is clearly a scalar.

This process is called *contracting* the indices, and it's very general. As long as A^μ is a 4-vector, the quantity $A^\mu A_\mu$ is a scalar. We can take any 4-vector at all and make a scalar by contracting its indices. We can also write $A^\mu A_\mu$ a little differently by referring to Eq. 5.20 and replacing A_μ with $\eta_{\mu\nu} A^\nu$. In other words, we can write

$$A^\mu A_\mu = A^\mu \eta_{\mu\nu} A^\nu. \qquad (5.21)$$

On the right side, we use the metric η and sum over μ and ν. Both sides of Eq. 5.21 represent the same scalar. Now let's look at an example involving two *different* 4-vectors, A and B. Consider the expression

$$A^\mu B_\mu.$$

Is this a scalar? It certainly looks like one. It has no indices because all the indices have been summed over.

To prove that it's a scalar, we'll need to rely on the fact that the sums and differences of scalars are also scalars. If we have two scalar quantities, then by definition you and I will agree about their values even though our reference frames are different. But if we agree about their values, we must also agree about the value of their sum and the value of their difference. Therefore, the sum of two scalars is a scalar, and the difference of two scalars is also a scalar. If we keep this in mind, the proof is easy. Just start with two 4-vectors A^μ and B^μ and write the expression

$$(A + B)^\mu (A + B)_\mu.$$

This expression must be a scalar. Why is that? Because both A^μ and B^μ are 4-vectors, their sum $(A + B)^\mu$ is also a 4-vector. If you contract any 4-vector with itself, the result is a scalar. Now, let's modify this expression by subtracting $(A - B)^\mu (A - B)_\mu$. This becomes

$$(A + B)^\mu (A + B)_\mu - (A - B)^\mu (A - B)_\mu. \tag{5.22}$$

This modified expression is still a scalar because it's the difference of two scalars. If we expand the expression, we find that the $A^\mu A_\mu$ terms cancel, and so do the $B^\mu B_\mu$ terms. The only remaining terms are $A^\mu B_\mu$ and $A_\mu B^\mu$, and the result is

$$2\left[A^\mu B_\mu + A_\mu B^\mu \right]. \tag{5.23}$$

I'll leave it as an exercise to prove that

$$A^\mu B_\mu = A_\mu B^\mu.$$

It doesn't matter if you put the ups down and the downs up;

the result is the same. Therefore, the expression evaluates to

$$(A + B)^{\mu}(A + B)_{\mu} - (A - B)^{\mu}(A - B)_{\mu} = 4\left[A^{\mu}B_{\mu}\right].$$

Because we know that the original Expression 5.22 is a scalar, the result $A^{\mu}B_{\mu}$ must also be a scalar.

You may have noticed that the expression $A^{\mu}B_{\mu}$ looks a lot like the ordinary dot product of two space vectors. You can think of $A^{\mu}B_{\mu}$ as the Lorentz or Minkowski version of the dot product. The only real difference is the change of sign for the time component, facilitated by the metric η.

5.5   Scalar Field Conventions

Next, we'll set up some conventions for a scalar field, $\phi(x)$. In this discussion, x represents all four components of space-time, including time. Before we get rolling, I need to state a theorem. The proof is not hard, and I'll leave it as an exercise.

Suppose you have a known 4-vector A_{μ}. To say A_{μ} is a 4-vector is not just a statement that it has four components. It means that A_{μ} transforms in a particular way under Lorentz transformations. Suppose also that we have another quantity B^{μ}. We don't know whether B^{μ} is a 4-vector. But we are told that when we form the expression

$$A_{\mu}B^{\mu},$$

the result is a scalar. Given these conditions, it's possible to prove that B^{μ} must be a 4-vector. With this result in mind, let's consider the change in value of $\phi(x)$ between two neighboring points. If $\phi(x)$ is a scalar, you and I will agree on its value at each of two neighboring points. Therefore, we'll also agree on the difference of its values at these two points: If $\phi(x)$ is a scalar, the change in $\phi(x)$ between two neighboring points is also a scalar.

What if the two neighboring points are infinitesimally separated? How do we express the difference in $\phi(x)$ between these neighboring points? The answer comes from basic calculus: Differentiate $\phi(x)$ with respect to each of the coordinates, and multiply that derivative by the differential of that coordinate,

$$d\phi(x) = \frac{\partial \phi(x)}{\partial X^{\mu}} dX^{\mu}. \qquad (5.24)$$

Following the summation convention, the right side is the derivative of $\phi(x)$ with respect to t times dt, plus the derivative of $\phi(x)$ with respect to x times dx, and so on. It's the small change in $\phi(x)$ when going from one point to another. We already know that both $\phi(x)$ and $d\phi(x)$ are scalars. Clearly, dX^{μ} itself is a 4-vector. In fact, it's the basic prototype of a 4-vector. To summarize: We know that the left side of Eq. 5.24 is a scalar, and that dX^{μ} on the right side is a 4-vector. What does that tell us about the partial derivative on the right side? According to our theorem, it must be a 4-vector. It stands for a complex of four quantities,

$$\frac{\partial \phi(x)}{\partial X^{\mu}} = \left(\frac{\partial \phi}{\partial t}, \frac{\partial \phi}{\partial x}, \frac{\partial \phi}{\partial y}, \frac{\partial \phi}{\partial z} \right).$$

For Eq. 5.24 to make sense as a product, the quantity $\dfrac{\partial \phi(x)}{\partial X^{\mu}}$ must correspond to a *covariant* vector, because the differential dX^{μ} is contravariant. We have discovered that derivatives of a scalar $\phi(x)$ with respect to the coordinates are the covariant components A_{μ} of a 4-vector. This is worth emphasizing: The derivatives of a scalar with respect to X^{μ} form a covariant vector. They're sometimes written with the shorthand symbol $\partial_{\mu}\phi$, which we now define to be

$$\partial_{\mu}\phi = \frac{\partial \phi(x)}{\partial X^{\mu}}.$$

The symbol $\partial_\mu \phi$ has a lower index to indicate that its components are covariant. Is there a contravariant version of this symbol? You bet. The contravariant version has nearly the same meaning except that its time component has the opposite sign. Let's write this out explicitly:

$$\partial_\mu \phi \iff \left(\frac{\partial \phi}{\partial t}, \frac{\partial \phi}{\partial x}, \frac{\partial \phi}{\partial y}, \frac{\partial \phi}{\partial z} \right)$$

$$\partial^\mu \phi \iff \left(-\frac{\partial \phi}{\partial t}, \frac{\partial \phi}{\partial x}, \frac{\partial \phi}{\partial y}, \frac{\partial \phi}{\partial z} \right).$$

5.6   A New Scalar

We now have the tools we need to construct a new scalar. The new scalar is

$$\partial^\mu \phi \partial_\mu \phi.$$

If we expand this using the summation convention, we get

$$\partial^\mu \phi \partial_\mu \phi = -\left(\frac{\partial \phi}{\partial t} \right)^2 + \left(\frac{\partial \phi}{\partial x} \right)^2 + \left(\frac{\partial \phi}{\partial y} \right)^2 + \left(\frac{\partial \phi}{\partial z} \right)^2.$$

What does this stand for? It's similar to a field Lagrangian that we wrote down earlier. Eq. 4.7 contains the same expression with the signs reversed.[10] In our new notation, the Lagrangian is

$$\mathcal{L} = -\frac{1}{2} \partial^\mu \phi \partial_\mu \phi.$$

This makes it easy to see that the Lagrangian for that scalar field is itself a scalar. As we explained before, having a

---

[10] Eq. 4.7 also contains a potential energy term $-V(\phi)$, which we're ignoring for now.

scalar-valued Lagrangian is critical because an invariant Lagrangian leads to equations of motion that are invariant in form. Much of field theory is about the construction of invariant Lagrangians. So far, scalars and 4-vectors have been our main ingredients. Going forward, we will need to construct scalar Lagrangians from other objects as well: things like spinors and tensors. The notation we've developed here will make that task much easier.

5.7 Transforming Covariant Components

The familiar Lorentz transformation equations, as presented, apply to contravariant components. The equations are slightly different for covariant components. Let's see how this works. The familiar contravariant transformations for t and x are

$$(A')^t = \frac{A^t - vA^x}{\sqrt{1 - v^2}}$$

$$(A')^x = \frac{A^x - vA^t}{\sqrt{1 - v^2}}.$$

The covariant components are the same, except for the time component. In other words, we can replace A^x with A_x, and $(A')^x$ with $(A')_x$. However, the covariant time component A_t is the *negative* of the contravariant time component. So we must replace A^t with $-(A)_t$, and $(A')^t$ with $-(A')_t$. Making these substitutions in the first equation results in

$$-(A')_t = \frac{-A_t - vA_x}{\sqrt{1 - v^2}},$$

which simplifies to

$$(A')_t = \frac{A_t + vA_x}{\sqrt{1 - v^2}}.$$

Applying them to the second equation gives us

$$(A')_x = \frac{A_x + vA_t}{\sqrt{1 - v^2}}.$$

These equations are almost the same as the contravariant versions, except that the sign of v has been reversed.

5.8 Mathematical Interlude: Using Exponentials to Solve Wave Equations

Starting with a known Lagrangian, the Euler-Lagrange equations provide a template for writing the equations of motion. The equations of motion are themselves differential equations. For some purposes, knowing the form of these equations is good enough. However, sometimes we would like to solve them.

Finding solutions to differential equations is a huge topic. Nevertheless, stripped down to its bare bones, the basic approach is as follows:

1. Propose (okay, *guess*) a function that might satisfy the differential equation.
2. Plug the function into the differential equation. If it works, you're done. Otherwise, return to Step 1.

Instead of racking our brains over a solution, we'll just provide one that happens to work. It turns out that exponential functions of the form

$$\phi(x) = e^{i(kx - \omega t)}$$

are the main building blocks for wave equations. You may find it puzzling that we choose a complex valued function as a solution, when our problems assume that ϕ is a real valued scalar field. To make sense of this, remember that

$$e^{i(kx - \omega t)} = \cos(kx - \omega t) + i\sin(kx - \omega t), \qquad (5.25)$$

where $kx - \omega t$ is real. Eq. 5.25 highlights the fact that a complex function is the sum of a real function and i times another real function. Once we've worked out our solution $e^{i(kx - \omega t)}$, we regard these two real functions as two solutions and ignore the i. This is easy to see if the complex function is set to zero. In that case, its real and imaginary parts must separately equal zero, and both parts are solutions.[11] In Eq. 5.25,

$$\cos(kx - \omega t)$$

is the real part, and

$$\sin(kx - \omega t)$$

is the imaginary part.

If we ultimately extract real functions as our solutions, why bother with complex functions at all? The reason is that it's easy to manipulate the derivatives of exponential functions.

---

[11] Confusingly, the imaginary part of a complex function is the *real* function that multiplies i.

5.9 Waves

Let's look at the wave equation and solve it. We already have
the Lagrangian for ϕ,

$$\mathcal{L} = \frac{1}{2}\left[\left(\frac{\partial \phi}{\partial t}\right)^2 - \left(\frac{\partial \phi}{\partial x}\right)^2 - \left(\frac{\partial \phi}{\partial y}\right)^2 - \left(\frac{\partial \phi}{\partial z}\right)^2\right].$$

However, I want to extend it slightly by adding one more
term. The additional term, $-\frac{1}{2}\mu^2\phi^2$, is also a scalar. It's a
simple function of ϕ and does not contain any derivatives.
The parameter μ^2 is a constant. Our modified field Lagrangian is

$$\mathcal{L} = \frac{1}{2}\left[\left(\frac{\partial \phi}{\partial t}\right)^2 - \left(\frac{\partial \phi}{\partial x}\right)^2 - \left(\frac{\partial \phi}{\partial y}\right)^2 - \left(\frac{\partial \phi}{\partial z}\right)^2 - \mu^2\phi^2\right].$$

$$(5.26)$$

This Lagrangian represents the field theory analog of the
harmonic oscillator. If we were discussing a harmonic oscillator and we called the coordinate of the oscillator ϕ, the
kinetic energy would be

$$\frac{\dot{\phi}^2}{2}.$$

The potential energy would be $\frac{1}{2}\mu^2\phi^2$, where μ^2 represents a
spring constant. The Lagrangian would be

$$\frac{\dot{\phi}^2}{2} - \frac{\mu^2\phi^2}{2}.$$

This Lagrangian would represent the good old harmonic
oscillator. It's similar to Eq. 5.26, our field Lagrangian. The
only difference is that the field Lagrangian has some space
derivatives. Let's work out the equations of motion that
correspond to Eq. 5.26 and then solve them. We'll start with

the time component. The Euler-Lagrange equations tell us to calculate

$$\frac{d}{dt}\frac{\partial \mathcal{L}}{\partial\left(\dfrac{\partial \phi}{\partial t}\right)}$$

for Eq. 5.26. It should be easy to see that the result is

$$\frac{d}{dt}\frac{\partial \mathcal{L}}{\partial\left(\dfrac{\partial \phi}{\partial t}\right)} = \frac{\partial^2 \phi}{\partial t^2}.$$

This is the analog of the acceleration term for the harmonic oscillator. We get additional terms by taking derivatives of the space components of Eq. 5.26. With these additional terms, the left side of the equation of motion becomes

$$\frac{\partial^2 \phi}{\partial t^2} - \frac{\partial^2 \phi}{\partial x^2} - \frac{\partial^2 \phi}{\partial y^2} - \frac{\partial^2 \phi}{\partial z^2}.$$

To find the right side, we calculate $\dfrac{\partial \mathcal{L}}{\partial \phi}$. The result of that calculation is

$$\frac{\partial \mathcal{L}}{\partial \phi} = -\mu^2 \phi.$$

Gathering our results for the left and right sides of the Euler-Lagrange equations, the equation of motion for ϕ is

$$\frac{\partial^2 \phi}{\partial t^2} - \frac{\partial^2 \phi}{\partial x^2} - \frac{\partial^2 \phi}{\partial y^2} - \frac{\partial^2 \phi}{\partial z^2} = -\mu^2 \phi.$$

Now let's put everything on the left side, giving us

$$\frac{\partial^2 \phi}{\partial t^2} - \frac{\partial^2 \phi}{\partial x^2} - \frac{\partial^2 \phi}{\partial y^2} - \frac{\partial^2 \phi}{\partial z^2} + \mu^2 \phi = 0. \tag{5.27}$$

This is a nice simple equation. Do you recognize it? It's the

Klein-Gordon equation. It preceded the Schrödinger equation and was an attempt to describe a quantum mechanical particle. The Schrödinger equation is similar.[12] Klein and Gordon made the mistake of trying to be relativistic. Had they not tried to be relativistic, they would have written the Schrödinger equation and become very famous. Instead, they wrote a relativistic equation and became much less famous. The Klein-Gordon equation's connection to quantum mechanics is not important for now. What we want to do is solve it.

There are many solutions, all of them built up from plane waves. When working with oscillating systems, it's useful to pretend that the coordinate is complex. Then, at the end of the calculation, we look at the real parts and ignore the i. We explained this idea in the preceding Mathematical Interlude.

The solutions that interest us are the ones that oscillate with time and have a component of the form

$$e^{-i\omega t}.$$

This function oscillates with frequency ω. But we're interested in solutions that also oscillate in space, which have the form e^{ikx}. In three dimensions, we can write this as

$$e^{i(k_x x + k_y y + k_z z)},$$

where the three numbers k_x, k_y, and k_z are called the wave numbers.[13] The product of these two functions,

$$\phi = e^{-i\omega t}e^{i(k_x x + k_y y + k_z z)}, \tag{5.28}$$

---

[12] The Schrödinger equation only has a first derivative with respect to time, and includes the value i.

[13] You can think of them as three components of a wave vector, where $\vec{k} \cdot \vec{x} = k_x x + k_y y + k_z z$.

is a function that oscillates in space *and* in time. We'll look for solutions of this form.

Incidentally, there's a slick way to express the right side of Eq. 5.28. We can write it as

$$e^{-i\omega t} e^{i(k_x x + k_y y + k_z z)} = e^{i(k_\mu X^\mu)}. \tag{5.29}$$

Where does that expression come from? If you think of k as a 4-vector, with components $(-\omega, k_x, k_y, k_z)$, then the expression $k_\mu X^\mu$ on the right side is just $-\omega t + k_x x + k_y y + k_z z$.[14] This notation is elegant, but for now we'll stick to the original form.

Let's see what happens if we try to plug our proposed solution (Eq. 5.28) into the equation of motion (Eq. 5.27). We'll be taking various derivatives of ϕ. Eq. 5.27 tells us which derivatives to take. We start by taking the second derivative of ϕ with respect to time. Differentiating Eq. 5.28 twice with respect to time gives us

$$\frac{\partial^2 \phi}{\partial t^2} = -\omega^2 \phi.$$

Differentiating twice with respect to x results in

$$\frac{\partial^2 \phi}{\partial x^2} = -k_x{}^2 \phi.$$

We get similar results when we differentiate with respect to y and z. So far, the Klein-Gordon equation has generated the terms

$$(-\omega^2 + k_x{}^2 + k_y{}^2 + k_z{}^2)\phi$$

based on our proposed solution. But we're not finished. Eq.

---

[14] It turns out that $(-\omega, k_x, k_y, k_z)$ really is a 4-vector, but we haven't proved it.

5.27 also contains the term $+\mu^2\phi$. This has to be added to the other terms. The result is

$$(-\omega^2 + k_x{}^2 + k_y{}^2 + k_z{}^2 + \mu^2)\phi = 0.$$

At this point, it's easy to find a solution. We just set the factor inside the parentheses equal to zero and find that

$$\omega = \pm\sqrt{k_x{}^2 + k_y{}^2 + k_z{}^2 + \mu^2}. \qquad (5.30)$$

This tells us the frequency in terms of the wave numbers. Either $+\omega$ or $-\omega$ will satisfy this equation. Also, notice that each term under the square root is itself a square. So if a particular value of (say) k_x is part of a solution, then its negative will also be a part of a solution.

Notice the parallel between these solutions and the energy equation, Eq. 3.43, from Lecture 3, repeated here:

$$E = \pm\sqrt{P^2 + m^2}.$$

Eq. 5.30 represents the classical field version of an equation that describes a quantum mechanical particle with mass μ, energy ω, and momentum k.[15] We'll come back to it again and again.

---

[15] The equation should also include some Planck's constants that I've ignored.

Interlude: Crazy Units

"Hi, Art, are you up for some talk about units?"

"Electromagnetic units? Oy vey, I'd rather eat wormholes. Do we have to?"

"Well, take your pick: units or a wormhole dinner."

"Okay, Lenny, you win—units."

When I first started learning about physics, something bothered me: Why are all the numbers—the so-called constants of nature—so big or so small? Newton's gravitational constant is 6.7×10^{-11}, Avogadro's number is 6.02×10^{23}, the speed of light is 3×10^8, Planck's constant is 6.6×10^{-34}, and the size of an atom is 10^{-10}. Nothing like this ever happened when I was learning mathematics. True, $\pi \approx 3.14159$ and $e \approx 2.718$. These natural mathematical numbers were neither big nor small, and although they had their own transcendental oddness, I could use the math I knew to work out their values. I understand why biology might have nasty numbers—it's a messy subject—but physics? Why such ugliness in the fundamental laws of nature?

I.1   Units and Scale

The answer turned out to be that the numerical values of the so-called constants of nature actually have more to do with biology than physics.[1] As an example, take the size of an atom, about 10^{-10} meters. But why do we measure in meters? Where did the meter come from and why is it so much bigger than an atom?

When asked this way, the answer starts to come into focus. A meter is simply a unit that's convenient for measuring ordinary human-scale lengths. It seems that the meter arose as a unit for measuring rope or cloth and it was simply the distance from a man's nose (supposedly the king's nose) to his outstretched fingertips.

But that raises the question, why is a man's arm so long—

---

[1] If you've read our previous book on quantum mechanics, you've heard this sermon before. It's interesting that the same issues of scale that affect our choice of units also limit our ability to directly perceive quantum effects with our senses.

10^{10}—in atomic radii? Here the answer is obvious: It takes a lot of atoms to make an intelligent creature that can even bother to measure rope. The smallness of an atom is really all about biology, not physics. You with me, Art?

And what about the speed of light; why so large? Here again the answer may have more to do with life than with physics. There are certainly places in the universe where things—even large, massive things—move relative to each other with speeds close to the speed of light. Only recently, two black holes were discovered to be orbiting each other at an appreciable fraction of the speed of light. They crashed into each other, but that's the way it goes; moving that fast can be dangerous. In fact, an environment full of objects whizzing around at nearly the speed of light would be lethal for our soft bodies. So the fact that light moves very fast on the scale of ordinary human experience is, at least in part, biology. We can only live where things with appreciable mass move slowly.

Avogadro's number? Again, intelligent creatures are necessarily big on the molecular scale, and the objects that we can easily handle, like beakers and test tubes, are also big. The quantities of gas and fluid that fill a beaker are large (in number of molecules) for reasons of convenience to our large soft selves.

Are there better units more suited to the fundamental principles of physics? Yes indeed, but let's first recall that in standard textbooks we are told that there are three fundamental units: length, mass, and time. If, for example, we chose to measure length in units of the radius of a hydrogen atom, instead of the length of a man's arm, there would be no large or small constants in the equations of atomic physics or chemistry.

But there is nothing universal about the radius of an atom. Nuclear physicists might still complain about the small size of the proton or, even worse, the size of the quark. The obvious fix would be to use the quark radius as the standard of length. But a quantum-gravity theorist would complain: "Look here, my equations are still ugly; the Planck length is 10^{-19} in your stupid nuclear physics units. Furthermore, the Planck length is much more fundamental than the size of a quark."

I.2   Planck Units

As the books say, there are three units: length, mass, and time. Is there a most natural set of units? To put it another way, are there three phenomena that are so fundamental, so universal, that we can use them to define the most fundamental choice of units? I think there are, and so did Planck in 1900. The idea is to pick aspects of physics that are completely universal, meaning that they apply with equal force to all physical systems. With some minor historical distortion, here was Planck's reasoning:

The first universal fact is that there is a speed limit that all matter must respect. No object—*no object*—can exceed the speed of light, whether it be a photon or a bowling ball. That gives the speed of light a universal aspect that the speed of sound, or any other speed, does not have. So Planck said, let's choose the most fundamental units so that the speed of light is 1, $c = 1$.

Next he said that gravity provides us with something universal, Newton's universal law of gravitation: *Every object in the universe attracts every other object with a force equal to Newton's constant times the product of the masses, divided by the square of the distance between them.* There are no exceptions; nothing is immune to gravity. Again, Planck recognized something universal about gravity that is not true of other forces. He concluded that the most fundamental choice of units should be defined so that Newton's gravitational constant is set to unity: $G = 1$.

Finally, a third universal fact of nature—one that Planck could not fully appreciate in 1900—is the Heisenberg Uncertainty Principle. Without too much explanation, what it says is that all objects in nature are subject to the same limitation on the accuracy with which they can be known: *The product of the uncertainty in position, and the uncertainty of momentum, is at least as big as Planck's constant divided by 2,*

$$\Delta x \Delta p \geq \frac{\hbar}{2}. \tag{I.1}$$

Again, this is a universal property that applies to every object, no matter how big or small—humans, atoms, quarks, and everything else. Planck's conclusion: The most fundamental units should be such that his own constant $\hbar$ is set to 1. This, as it turns out, is enough to fix the three basic units of length, mass, and time. Today, the resulting units are called Planck units.

So then why don't all physicists use Planck units? There is no doubt that the fundamental laws of physics would be most simply expressed. In fact, many theoretical physicists do work in Planck units, but they would not be at all convenient for ordinary purposes. Imagine if we used Planck units in daily life. The signs on freeways would read

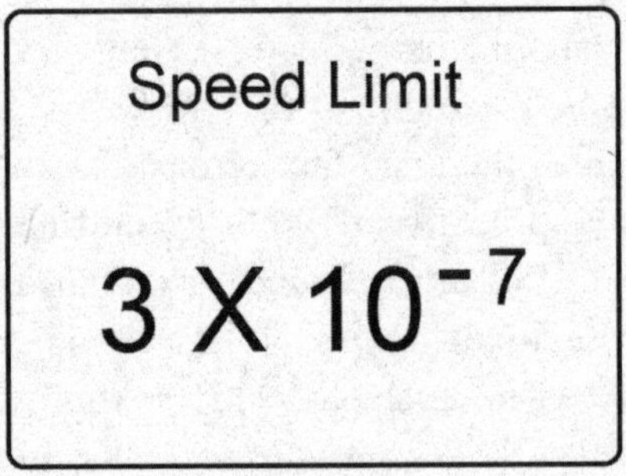

Figure I.1: Traffic Sign in Planck Units.

The distance to the next exit would be 10^{38}, and the time in a single day would be 8.6×10^{46}. Perhaps more important for physics, ordinary laboratory units would have inconveniently large and small values. So for convenience's sake we live with units that are tailored to our biological limitations. (By the way, none of this explains the incredible fact that in our country we still measure in inches, feet, yards, slugs, pints, quarts, and teaspoons.)

I.3 Electromagnetic Units

Art: *Okay, Lenny, I get what you are saying. But what about electromagnetic units? They seem to be especially annoying. What's that thing ϵ_0 in all the equations, the thing the textbooks call the dielectric constant of the vacuum?[2] Why does the vacuum have a dielectric constant anyway, and why is it equal to 8.85×10^{-12}? That seems really weird.*

Art is right; electromagnetic units are a nuisance all of their own. And he is right that it doesn't make sense to think of

---

[2] Also called vacuum permittivity, or permittivity of free space.

the vacuum as a dielectric—not in classical physics, anyway. The language is a holdover from the old ether theory.

The real question is: Why was it necessary to introduce a new unit for electric charge—the so-called coulomb? The history is interesting and actually based on some physical facts, but probably not the ones you imagine. I'll start by telling you how I would have set things up, and why it would have failed.

What I would have done is to start by trying to accurately measure the force between two electric charges, let's say by rubbing two pith balls with cat's fur until they were charged. Presumably I would have found that the force was governed by Coulomb's law,

$$F = \frac{q_1 q_2}{r^2}. \tag{I.2}$$

Then I would have declared that a unit charge—one for which $q = 1$—is an amount of charge such that two of them, separated by 1 meter, have a force between them of 1 newton. (The newton is a unit of force needed to accelerate a 1-kilogram mass by 1 meter per second per second.) In that way there would be no need for a new independent unit of charge, and Coulomb's law would be simple, just like I wrote earlier.

Maybe if I had been particularly clever and had a bit of foresight, I might have put a factor of 4π in the denominator of the Coulomb law:

$$F = \frac{q_1 q_2}{4\pi r^2}. \tag{I.3}$$

But that's a detail.

Now why would I have failed, or at least not had good accuracy? The reason is that it is difficult to work with charges; they are hard to control. Putting a decent amount of

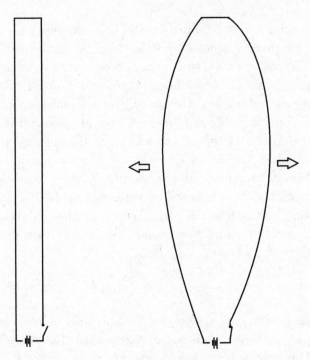

Figure I.2: Parallel
Wires. No current
flows because the
switch is open.

Figure I.3: Parallel Wires with
Currents Flowing in Opposite
Directions. The switch is closed.

charge on a pith ball is hard because the electrons repel and
tend to jump off the ball. So historically a different strategy
was used.

By contrast with charge, working with electric current in
wires is easy. Current is charge in motion, but because the
negative charge of the moving electrons in a wire is held in
place by the positive charges of the nuclei, they are easy to
control. So instead of measuring the force between two static
charges, we instead measure the force between two current-
carrying wires. Figures I.2 and I.3 illustrate how such an
apparatus might work. We start with a circuit containing a

battery, a switch, and two long parallel wires stretched tight and separated by a known distance. For simplicity the distance could be a meter, although in practice we may want it to be a good deal smaller.

Now we close the switch and let the current flow. The wires repel each other for reasons that will be explained later in this book. What we see is that the wires belly out in response to the force. In fact, we can use the amount of bellying to measure the force (per unit length). This allows us to define a unit of electric current called an *ampere* or *amp.*

One amp is the current needed to cause parallel wires separated by 1 meter, to repel with a force of 1 newton per meter of length.

Notice that in this way, we define a unit of current, not a unit of electric charge. Current measures the amount of charge passing any point on the circuit per unit time. For example, it is the amount of charge that passes through the battery in 1 second.

Art: *But wait, Lenny. Doesn't that also allow us to define a unit of charge? Can't we say that our unit of charge—call it 1 coulomb of charge—is the amount of charge that passes through the battery in one second, given that the current is one amp?*

Lenny: *Very good! That's exactly right. Let me say it again: The coulomb is by definition the amount of charge passing through the circuit in one second, when the current is one amp; that is, when the force on the wires is one newton per meter of length (assuming the wires are separated by one meter).*

The disadvantage is that the definition of a coulomb is indirect. The advantage is that the experiment is so easy that even I did it in the lab. The problem, however, is that the unit of charge defined this way is not the same unit that would result from measuring the force between static charges.

How do the units compare? To answer that, we might try to collect two buckets of charge, each a coulomb, and measure the force between them. This would be dangerous even if it were possible; a coulomb is a really huge amount of charge. The bucket would explode and the charge would just fly apart. So the question becomes, why does it take such a huge amount of charge flowing in the wires to produce a modest force of 1 newton?

Art: *Why is there a force between the wires anyway? Even though the wires have moving electrons, the net charge on the wire is zero. I don't see why there is any force.*

Lenny: *Yes, you are right that the net charge is zero. The force is not electrostatic. It's actually due to the magnetic field caused by the motion of the charges. The positive nuclei are at rest and don't cause a magnetic field, but the moving electrons do.*

Art: *Okay, but you still haven't told me why it takes such a whopping big amount of moving charge to create a mere newton of force between the wires. Am I missing something?*

Lenny: *Only one thing. The charges move very slowly.*

The electrons in a typical current-carrying wire do indeed move very slowly. They bounce around very quickly, but like a drunken sailor, they mostly get nowhere; on the average it takes an electron about an hour to move 1 meter along a wire. That seems slow, but compared to what? The answer is that they move very slowly compared to the only natural physical unit of velocity, the speed of light. In the end that's why it takes a huge amount of charge, moving through the wires of a circuit, to produce a significant force.

Now that we know that the standard unit of charge, the

coulomb, is a tremendously large amount of charge, let's go
back to Coulomb's law. The force between two coulomb-size
charges is enormous. To account for this we have to put a
huge constant into the force law. Instead of

$$F = \frac{q_1 q_2}{4\pi r^2},$$

we write

$$F = \frac{q_1 q_2}{4\pi\epsilon_0 r^2}, \tag{I.4}$$

where ϵ_0 is the small number 8.85×10^{-12}.

Art: *So ultimately the weird dielectric constant of the
vacuum has nothing to do with dielectrics. It has more to do
with the slow molasses-like motion of electrons in metallic
wires. Why don't we just get rid of ϵ_0 and set it equal to 1?*

Lenny: *Good idea, Art. Let's do that from now on. But
don't forget that we will be working with a unit of charge that
is about one three-hundred-thousandth of a Coulomb. Forgetting about the conversion factor could lead to a nasty
explosion.*

Lecture 6

The Lorentz Force Law[1]

Art: *Lenny, who is that dignified gentleman with the beard and wire-frame glasses?*

Lenny: *Ah, the Dutch uncle. That's Hendrik. Would you like to meet him?*

Art: *Sure, is he a friend of yours?*

Lenny: *Art, they're all friends of mine. Come on, I'll introduce you.*

Art Friedman, meet my friend Hendrik Lorentz.

Poor Art, he's not quite prepared for this.

Art: *Lorentz? Did you say Lorentz? Oh my God! Are you? Is he? Are you really, are you really the ..."*

Dignified as always, HL bows deeply.

Lorentz: *Hendrik Antoon Lorentz, at your service.*

---

[1] Since these dialogues take place in an alternate universe, Art gets to meet Lorentz for the first time—again!

Later a star-struck Art quietly asks, *Lenny, is that really* the *Lorentz? The one who discovered Lorentz transformations?*

Lenny: *Sure he is, and a lot more than that. Bring me a napkin and a pen and I'll tell you about his force law.*

Of all the fundamental forces in nature, and there are many of them, few were known before the 1930s. Most are deeply hidden in the microscopic quantum world and only became observable with the advent of modern elementary particle physics. Most of the fundamental forces are what physicists call *short-range*. This means that they act only on objects that are separated by very small distances. The influence of a short-range force decreases so rapidly when the objects are separated that, for the most part, they are not noticed in the ordinary world. An example is the so-called nuclear force between nucleons (protons and neutrons). It's a powerful force whose role is to bind these particles into nuclei. But as powerful as that force is, we don't ordinarily notice it. The reason is that its effects disappear exponentially when the nucleons are separated by more than about 10^{-15} meters. The forces that we do notice are the *long-range* forces whose effects fade slowly with distance.

Of all the forces of nature, only three were known to the ancients—electric, magnetic, and gravitational. Thales of Miletos (600 BC) was said to have moved feathers with amber that had been rubbed with cat fur. At about the same time he mentioned lodestone, a naturally occurring magnetic material. Aristotle, who was probably late on the scene, had a theory of gravity, even if it was completely wrong. These three were the only forces that were known until the 1930s.

What makes these easily observed forces special is that they are *long-range*. Long-range forces fade slowly with distance and can be seen between objects when they are well separated.

Gravitational force is by far the most obvious of the three, but surprisingly it is much weaker than electromagnetic force. The reason is interesting and worth a short digression. It goes back to Newton's universal law of gravitational attraction: Everything attracts everything else. Every elementary particle in your body is attracted by every particle in the Earth. That's a lot of particles, all attracting one another, and the result is a significant and noticeable gravitational attraction, but in fact the gravitational attraction between individual particles is far too small to measure.

The electric forces between charged particles is many orders of magnitude stronger than the gravitational force. But unlike gravity, electric force can be either attractive (between opposite charges) or repulsive (between like charges). Both you and the Earth are composed of an equal number of positive charges (protons) and negative charges (electrons), and the result is that the forces cancel. If we imagined getting rid of all the electrons in both you and the Earth, the repulsive electric forces would easily overwhelm gravity and blast you from the Earth's surface. In fact, it would be enough to blast the Earth and you into smithereens.

In any case, gravity is not the subject of these lectures, and the only other long-range forces are electromagnetic. Electric and magnetic forces are closely related to each other; in a sense they are a single thing, and the unifying link is relativity. As we will see, an electric force in one frame of reference becomes a magnetic force in another frame, and vice versa. To put it another way, electric and magnetic forces transform into each other under Lorentz transformation. The rest of these lectures are about electromagnetic forces and how they are unified into a single phenomenon through relativity. Going back to Pauli's (fictitious) paraphrase of John Wheeler's (real) slogan,

Fields tell charges how to move; charges tell fields how to vary.

We'll begin with the first half—fields tell charges how to

move. Or to put it more prosaically, fields determine the forces on charged particles.

An example that may be familiar to you—if not, it soon will be—is the electric field $\vec{E}$. Unlike the scalar field that we discussed in the last lecture, the electric field is a vector field—a 3-vector, to be precise. It has three components and points in some direction in space. It controls the electric force on a charged particle according to the equation

$$F = e\vec{E}.$$

In this equation the symbol e represents the electric charge of the particle. It can be positive, in which case the force is in the same direction as the electric field; it can be negative, in which case the force and the field are in opposite directions; or, as in the case of a neutral atom, the force can be zero.

Magnetic forces were first discovered by their action on magnets or bits of lodestone, but they also act on electrically charged particles if the particles are in motion. The formula involves a magnetic field $\vec{B}$ (also a 3-vector), the electric charge e, and the velocity of the particle $\vec{v}$. We'll derive it later from an action principle, but jumping ahead, the force on a charged particle due to a magnetic field is

$$F = e \, \vec{v} \times \vec{B}.$$

The symbol $\times$ represents the ordinary cross-product of vector algebra, which I assume you have seen before.[2] One interesting property of magnetic forces is that they vanish for a particle at rest and increase as the particle velocity increases. If there happens to be both an electric and a magnetic field, the full force is the sum

$$F = e(\vec{E} + \vec{v} \times \vec{B}). \tag{6.1}$$

___
[2] There's a brief review in Appendix B.

Eq. 6.1 was discovered by Lorentz and is called the *Lorentz force law.*

We've already discussed scalar fields and the way they interact with particles. We showed how the same Lagrangian (and the same action) that tells the field how to influence the particle also tells the particle how to influence the field. Going forward, we'll do the same thing for charged particles and the electromagnetic field. But before we do, I want to briefly review our notational scheme and extend it to include a new kind of object: *tensors.* Tensors are a generalization of vectors and scalars and include them as special cases. As we will see, the electric and magnetic fields are not separate entities but combine together to form a relativistic tensor.

6.1 Extending Our Notation

Our basic building blocks are 4-vectors with upper and lower indices. In the context of special relativity, there's little difference between the two types of indices. The only difference occurs in the time component (the component with index zero) of a 4-vector. For a given 4-vector, the time component with an upper index has the opposite sign of the time component with lower index. In symbols,

$$A^0 = -A_0.$$

It may seem like overkill to define a notation whose sole purpose is to keep track of sign changes for time components. However, this simple relationship is a special case of a much broader *geometric* relationship based on the metric tensor. When we study general relativity, the relationship between upper and lower indices will become far more interesting. For now, upper and lower indices simply provide a convenient, elegant, and compact way to write equations.

6.1.1   4-Vector Summary

Here is a quick summary of concepts from Lecture 5 for easy reference.

4-vectors have three space components and one time component. A Greek index such as μ refers to any or all of these components and may take the values $(0, 1, 2, 3)$. The first of these (the component labeled zero) is the time component. Components 1, 2, and 3 correspond to the x, y, and z directions of space. For example, A^0 represents the time component of the 4-vector A. A^2 represents the space component in the y direction. When we focus only on the three space components of a 4-vector, we label them with Latin indices such as m or p. Latin indices may take the values $(1, 2, 3)$, but not zero. Symbolically, we can write

$$A^\mu \longrightarrow A^0, A^m.$$

So far, I've labeled my 4-vectors with upstairs, or *contravariant* indices. By convention, a contravariant index is the sort of thing you would attach to the coordinates themselves, such as X^μ, or to a coordinate displacement such as dX^μ. Things that transform in the same way as coordinates or displacements carry upstairs indices.

The covariant counterpart to A^μ is written with a downstairs index, A_μ. It describes the *same* 4-vector using a different notation. To switch from contravariant notation to covariant notation, we use the 4×4 metric

$$\eta_{\mu\nu} = \begin{pmatrix} -1 & 0 & 0 & 0 \\ 0 & 1 & 0 & 0 \\ 0 & 0 & 1 & 0 \\ 0 & 0 & 0 & 1 \end{pmatrix}.$$

The formula

$$A_\mu = \eta_{\mu\nu} A^\nu \tag{6.2}$$

converts an upstairs index to a downstairs index. The repeated index ν on the right side is a summation index, and therefore Eq. 6.2 is shorthand for

$$A_\mu = \eta_{\mu 0} A^0 + \eta_{\mu 1} A^1 + \eta_{\mu 2} A^2 + \eta_{\mu 3} A^3.$$

The covariant and contravariant components of any 4-vector A are exactly the same, except for the time components. The upstairs and downstairs time components have opposite signs. Eq. 6.2 is equivalent to the two equations

$$A_0 = -A^0$$

$$A_m = A^m.$$

Eq. 6.2 gives this same result because of the -1 in the upper-left position of $\eta_{\mu\nu}$.

6.1.2 Forming Scalars

For any two 4-vectors A and B, we can form a product $A_\nu B^\nu$ using the upstairs components of one and the downstairs components of the other.[3] The result is a scalar, which means it has the same value in every reference frame. In symbols,

$$A_\nu B^\nu = scalar.$$

The repeated index ν indicates a summation over four values. The long form of this expression is

$$A_0 B^0 + A_1 B^1 + A_2 B^2 + A_3 B^3 = scalar.$$

---

[3] It's okay if A and B happen to be the same vector.

6.1.3   Derivatives

Coordinates and their displacements are the prototypes for contravariant (upstairs) components. In the same way, derivatives are the prototype for covariant (downstairs) components. The symbol ∂_μ stands for

$$\frac{\partial}{\partial X^\mu}.$$

In Lecture 5, we explained why these four derivatives are the covariant components of a 4-vector. We can also write them in contravariant form. To summarize:

Covariant Components:

$$\partial_\mu \Longrightarrow \left(\frac{\partial}{\partial X^0}, \frac{\partial}{\partial X^1}, \frac{\partial}{\partial X^2}, \frac{\partial}{\partial X^3} \right).$$

Contravariant Components:

$$\partial^\mu \Longrightarrow \left(-\frac{\partial}{\partial X^0}, \frac{\partial}{\partial X^1}, \frac{\partial}{\partial X^2}, \frac{\partial}{\partial X^3} \right).$$

As usual, the only difference between them is the sign of the time component.

The symbol ∂_μ doesn't mean much all by itself; it has to act on some object. When it does, it adds a new index μ to that object. For example, if ∂_μ operates on a scalar, it creates a new object with *covariant* index μ. Taking a scalar field ϕ as a concrete example, we could write

$$\partial_\mu \phi = \frac{\partial \phi}{\partial X^\mu}.$$

The right side is a collection of derivatives that forms the covariant vector

$$\left(\frac{\partial \phi}{\partial X^0}, \frac{\partial \phi}{\partial X^1}, \frac{\partial \phi}{\partial X^2}, \frac{\partial \phi}{\partial X^3} \right).$$

The symbol ∂_μ also provides a new way to construct a scalar from a vector. Suppose we have a 4-vector $B^\mu(t, x)$ that depends on time and position. In other words B is a 4-vector field. If B is differentiable, it makes sense to consider the quantity

$$\partial_\mu B^\mu(t, x).$$

Under the summation convention, this expression tells us to differentiate B^μ with respect to each of the four components of spacetime and add up the results:

$$\partial_\mu B^\mu(t, x) = \frac{\partial B^0}{\partial X^0} + \frac{\partial B^1}{\partial X^1} + \frac{\partial B^2}{\partial X^2} + \frac{\partial B^3}{\partial X^3}.$$

The result is a scalar.

The summing process we've illustrated here is very general; it's called *index* contraction. Index contraction means identifying an upper index with an identical lower index within a single term, and then summing.

6.1.4 General Lorentz Transformation

Back in Lecture 1, we introduced the general Lorentz transformation. Here, we return to that idea and add some details.

Lorentz transformations make just as much sense along the y axis or the z axis as they do along the x axis. There's certainly nothing special about the x direction or any other direction. In Lecture 1, we explained that there's another class of transformations—rotations of space—that are also considered members of the family of Lorentz transformations. Rotations of space do not affect time components in any way.

Once you accept this broader definition of Lorentz invariance, you can say that a Lorentz transformation along the y axis is simply a rotation of the Lorentz transformation along the x axis. You can combine rotations together with the "normal" Lorentz transformations to make a Lorentz transformation in any direction or a rotation about any axis. This is the general set of transformations under which physics is invariant. The proof of this result is not important to us right now. What is important is that physics is invariant not only under simple Lorentz transformations but also under a broader category of transformations that includes rotations of space.

How can we fold Lorentz transformations into our index-based notation scheme? Let's consider the transformation of a contravariant vector

$$A^\mu.$$

By definition, this vector transforms in the same way as the contravariant displacement vector X^μ. For example, the transformation equation for the time component A^0 is

$$(A')^0 = \frac{A^0 - vA^1}{\sqrt{1 - v^2}}.$$

This is the familiar Lorentz transformation along the x axis except that I've called the time component A^0 and I've called the x component A^1. We can always write these transformations in the form of a matrix acting on the components of a vector. For example, I can write

$$(A')^\mu$$

to represent the components of the 4-vector A^μ in my frame of reference. To express these components as functions of the components in your frame, we'll define a matrix $L^\mu{}_\nu$ with upper index μ and lower index ν,

$$(A')^\mu = L^\mu{}_\nu A^\nu. \tag{6.3}$$

$L^{\mu}{}_{\nu}$ is a matrix because it has two indices; it's a 4 × 4 matrix that multiplies the 4-vector A^{ν}.[4]

Let's make sure that Eq. 6.3 is properly formed. The left side has a free index μ, which can take any of the values $(0, 1, 2, 3)$. The right side has two indices, μ and ν. The summation index ν is not an explicit variable in the equation. The only free index on the right side is μ. In other words, each side of the equation has a free contravariant index μ. Therefore, the equation is properly formed; it has the same number of free indices on the left side as it does on the right side, and their contravariant characters match.

Eq. 6.4 gives an example of how we would use $L^{\mu}{}_{\nu}$ in practice. We have filled in matrix elements that correspond to a Lorentz transformation along the x axis.[5]

$$
\begin{pmatrix} t' \\ x' \\ y' \\ z' \end{pmatrix} = \begin{pmatrix} \dfrac{1}{\sqrt{1-v^2}} & \dfrac{-v}{\sqrt{1-v^2}} & 0 & 0 \\ \dfrac{-v}{\sqrt{1-v^2}} & \dfrac{1}{\sqrt{1-v^2}} & 0 & 0 \\ 0 & 0 & 1 & 0 \\ 0 & 0 & 0 & 1 \end{pmatrix} \begin{pmatrix} t \\ x \\ y \\ z \end{pmatrix}. \tag{6.4}
$$

What does the equation say? Following the rules of matrix multiplication, Eq. 6.4 is equivalent to four simple equations.

---

[4] This equation exposes a slight conflict of notational conventions. Our convention for the use of Greek indices states that they range in value from 0 to 3. However, standard matrix notation assumes that index values run from 1 to 4. We will favor the 4-vector convention (0 to 3). This does no harm as long as we stick to the ordering (t, x, y, z).

[5] We could have labeled the components of the column vector (A^0, A^1, A^3, A^3) instead of (t, x, y, z), and ditto for the vector with primed components. These are just different labels that represent the same quantities.

The first equation specifies the value of t', which is the first element of the vector on the left side. We set t' equal to the dot product of the first row of the matrix with the column vector on the right side. In other words, the equation for t' becomes

$$t' = \frac{t}{\sqrt{1 - v^2}} - \frac{vx}{\sqrt{1 - v^2}} + 0y + 0z,$$

or simply

$$t' = \frac{t}{\sqrt{1 - v^2}} - \frac{vx}{\sqrt{1 - v^2}}.$$

Carrying out the same process for the second row of $L^\mu{}_\nu$ gives the equation for x'

$$x' = \frac{-vt}{\sqrt{1 - v^2}} + \frac{x}{\sqrt{1 - v^2}}.$$

The third and fourth rows produce the equations

$$y' = y$$

$$z' = z.$$

It's easy to recognize these equations as the standard Lorentz transformation along the x axis. If we wanted to transform along the y axis instead, we would just shuffle these matrix elements around. I'll leave it to you to figure out how to do that.

Now let's consider a different operation: a rotation in the y, z plane, where the variables t and x play no part at all. A rotation can also be represented as a matrix, but the elements would be different from our first example. To begin with, the upper-left quadrant would look like a 2×2 unit matrix. That assures that t and x are not affected by the transformation.

What about the lower-right quadrant? You probably know the answer. To rotate the coordinates by angle θ, the matrix elements would be the sines and cosines shown in Eq. 6.5.

$$
\begin{pmatrix} t' \\ x' \\ y' \\ z' \end{pmatrix} = \begin{pmatrix} 1 & 0 & 0 & 0 \\ 0 & 1 & 0 & 0 \\ 0 & 0 & \cos(\theta) & \sin(\theta) \\ 0 & 0 & -\sin(\theta) & \cos(\theta) \end{pmatrix} \begin{pmatrix} t \\ x \\ y \\ z \end{pmatrix} \tag{6.5}
$$

Following the rules of matrix multiplication, Eq. 6.5 is equivalent to the four equations

$$
\begin{aligned}
t' &= t \\
x' &= x \\
y' &= y\ \cos(\theta) + z\ \sin(\theta) \\
z' &= -y\ \sin(\theta) + z\ \cos(\theta).
\end{aligned} \tag{6.6}
$$

In a similar way, we can write matrices representing rotations in the x, y or x, z planes by shuffling these matrix elements to different locations within the matrix.

Once we define a set of transformation matrices for simple linear motion and spatial rotations, we can multiply these matrices together to make more complicated transformations. In this way, we can represent a complicated transformation using a single matrix. The simple transformation matrices shown here, along with their y and z counterparts, are the basic building blocks.

6.1.5 Covariant Transformations

So far, we've explained how to transform a 4-vector with a *contravariant* (upper) index. How do we transform a 4-vector with a covariant (lower) index?

Suppose you have a 4-vector with covariant components. You know what these components look like in your frame, and you want to find out what they look like in my frame. We need to define a new matrix $M_\mu{}^\nu$ such that

$$(A')_\mu = M_\mu{}^\nu A_\nu. \qquad (6.7)$$

This new matrix needs to have a lower index μ, because the resulting 4-vector on the left side has a lower index μ.

Remember that M represents the same Lorentz transformation as our contravariant transformation matrix L; the two matrices represent the same physical transformation between coordinate frames. Therefore M and L must be connected. Their connection is given by a simple matrix formula:

$$M = \eta L \eta.$$

I'll let you prove this on your own. We're not going to use M very much, but it's nice to know how L and M are connected for a given Lorentz transformation.

It turns out that η is its own inverse, just like the unit matrix is its own inverse. In symbols,

$$\eta^{-1} = \eta.$$

The reason is that each diagonal entry is its own inverse; the inverse of 1 is 1, and the inverse of -1 is -1.

Exercise 6.1: Given the transformation equation (Eq. 6.3) for the contravariant components of a 4-vector A^ν, where $L^\mu{}_\nu$ is a Lorentz transformation matrix, show that the Lorentz transformation for A's covariant components is

$$(A')_\mu = M_\mu{}^\nu A_\nu,$$

where

$$M = \eta L \eta.$$

6.2 Tensors

A tensor is a mathematical object that generalizes the notions of a scalar and a vector; in fact, scalars and vectors are examples of tensors. We will use tensors extensively.

6.2.1 Rank 2 Tensors

A simple way to approach tensors is think of them as "a thing with some number of indices." The number of indices a tensor has—the *rank* of the tensor—is an important characteristic. For example, a scalar is a tensor of rank zero (it has zero indices), and a 4-vector is a tensor of rank one. A tensor of rank two is a thing with two indices.

Let's look at a simple example. Consider two 4-vectors, A and B. As we've seen before, we can form a product from them by contraction, $A_\mu B^\mu$, where the result is a scalar. But now let's consider a more general kind of product, a product whose *result* has two indices, μ and ν. We'll start with the contravariant version. We just multiply A^μ and B^ν for each μ and ν:

$$A^\mu B^\nu.$$

How many components does this object have? Each index can take four different values, so $A^\mu B^\nu$ can take 4×4 or sixteen different values. We can list them: $A^0 B^0$, $A^0 B^1$, $A^0 B^2$, $A^0 B^3$, $A^1 B^0$, and so forth. The symbol $A^\mu B^\nu$ stands for a complex of sixteen different numbers. It's just the set of numbers you get by multiplying the μ component of A with the ν component of B. This object is a tensor of rank two. I'll use the generic label T for tensor:

$$T^{\mu\nu} = A^\mu B^\nu.$$

Not all tensors are constructed from two vectors in this way, but two vectors can always define a tensor. How does $T^{\mu\nu}$ transform? If we know how A transforms and we know how B transforms (which is the same way) we can immediately figure out the transformed components of $A^\mu B^\nu$, which we call

$$(T')^{\mu\nu} = (A')^\mu (B')^\nu.$$

But, of course, we do know how A and B transform; Eq. 6.3 tells us how. Substituting Eq. 6.3 into the right side for both A' and B' gives the result

$$(T')^{\mu\nu} = L^\mu{}_\sigma A^\sigma L^\nu{}_\tau B^\tau.$$

This requires some explanation. In Eq. 6.3, the repeated index was called ν. However, in the preceding equation, we replaced ν with σ for the A factor, and with τ for the B factor. Remember, these are summation indices and it doesn't matter what we call them as long as we're consistent. To avoid confusion, we want to keep the summation indices distinct from other indices and distinct from each other.

Each of the four symbols on the right side stands for a number. Therefore, we're allowed to reorder them. Let's group the matrix elements together:

$$(T')^{\mu\nu} = L^\mu{}_\sigma L^\nu{}_\tau A^\sigma B^\tau. \tag{6.8}$$

Notice that we can now identify $A^\sigma B^\tau$ on the right side with the *untransformed* tensor element $T^{\sigma\tau}$ and rewrite the equation as

$$(T')^{\mu\nu} = L^\mu{}_\sigma L^\nu{}_\tau T^{\sigma\tau}. \tag{6.9}$$

Eq. 6.9 gives us a new transformation rule about a new kind of object with two indices. $T^{\mu\nu}$ transforms with the action of the Lorentz matrix on each one of the indices.

6.2.2   Tensors of Higher Rank

We can invent more complicated kinds of tensors—tensors with three indices, for example, such as

$$T^{\mu\nu\lambda}.$$

How would this transform? You could think of it as the product of three 4-vectors with indices μ, ν, and λ. The transformation formula is straightforward:

$$(T')^{\mu\nu\lambda} = L^\mu{}_\sigma L^\nu{}_\tau L^\lambda{}_\kappa T^{\sigma\tau\kappa}. \tag{6.10}$$

There's a transformation matrix for each index, and we can generalize this pattern to any number of indices.

At the beginning of the section, I said that a tensor is a thing that has a bunch of indices. That's true, but it's not the whole story; not every object with a bunch of indices is considered a tensor. To qualify as a tensor the indexed object has to transform like the examples we've shown. The transformation formula may be modified for different numbers of indices, or to accommodate covariant indices, but it must follow this general pattern.

Although tensors are defined by these transformation properties, not every tensor is constructed by taking a product of two 4-vectors. For example, suppose we have two other 4-vectors, C and D. We could take the product $A^\mu B^\nu$ and add it to the product $C^\mu D^\nu$,

$$A^\mu B^\nu + C^\mu D^\nu.$$

Adding tensors produces other tensors, and the preceding expression is indeed a tensor. But it cannot in general be written as a product of two 4-vectors. Its tensor character is determined by its transformation properties, not by the fact that it may or may not be constructed from a pair of 4-vectors.

6.2.3 Invariance of Tensor Equations

Tensor notation is elegant and compact. But the real power behind it is that tensor equations are frame invariant. If two tensors are equal in one frame, they're equal in every frame. That's easy to prove, but keep in mind that for two tensors to be equal, *all* of their corresponding components must be equal. If every component of a tensor is equal to the corresponding component of some other tensor, then of course they're the same tensor.

Another way to say it is that if all the components of a tensor are zero in one reference frame, then they're zero in every frame. If a *single* component is zero in one frame, it may well be nonzero in a different frame. But if *all* the components are zero, they must be zero in every frame.

6.2.4   Raising and Lowering Indices

I've explained how to transform a tensor with all of its indices upstairs (contravariant components). I could write down the rules for transforming tensors with mixed upstairs and downstairs indices. Instead I'll just tell you that once you know how a tensor transforms, you can immediately deduce how its other variants transform. By *other variants* I mean different versions of the same tensor—the same geometric quantity—but with some of its upstairs indices pushed downstairs and vice versa. For example, consider the tensor

$$T^{\mu}{}_{\nu} = A^{\mu}B_{\nu}.$$

This is a tensor with one index upstairs and one index downstairs. How does it transform? *Never mind!* You don't need to worry about it because you already know how to transform $T^{\mu\sigma}$ (with both indices upstairs), and you also know how to raise and lower indices using the matrix η. Recall that

$$T^{\mu}{}_{\nu} = T^{\mu\sigma}\eta_{\sigma\nu}.$$

You can lower a tensor index in exactly the same way as you lower a 4-vector index. Just multiply by η as above, using the summation convention. The resulting object is $T^{\mu}{}_{\nu}$.

But there's an even easier way to think about it. Given a tensor with all of its indices upstairs, how do you pull some of them downstairs? It's simple. If the index that you're lowering is a time index, you multiply by -1. If it's a space index, you don't multiply at all. That's what η does. For example, the tensor component T^{00} is exactly the same as T_{00} because I've lowered *two* time indices,

$$T^{00} = T_{00}.$$

Lowering two time indices means multiplying the component

by -1 twice. It's like the relationship between

$$A^0 B^0$$

and

$$A_0 B_0.$$

There are two minus signs in going from A^0 to A_0, and from B^0 to B_0. Each lowered index introduces a minus sign, and the first product is equal to the second. But how does

$$A^0 B^1$$

compare with

$$A_0 B_1 \ ?$$

B^1 and B_1 are the same, because index 1 refers to a space component. But A^0 and A_0 have opposite signs because index 0 is a time component. The same relationship holds between tensor components T^{01} and T_{01}:

$$T^{01} = -T_{01},$$

because only one time component was lowered. Whenever you lower or raise a time component, you change signs. That's all there is to it.

6.2.5 Symmetric and Antisymmetric Tensors

In general, the tensor component

$$T^{\mu\nu}$$

is not the same as

$$T^{\nu\mu},$$

where the indices μ and ν have been reversed. Changing the order of the indices makes a difference. To illustrate this, consider the product of two 4-vectors A and B. It should be clear that

$$A^\mu B^\nu \neq A^\nu B^\mu.$$

For example, if we choose specific components such as 0 and 1, we can see that

$$A^0 B^1 \neq A^1 B^0.$$

Clearly, these are not always the same: A and B are two different 4-vectors and there's no built-in reason for any of the components A^0, A^1, B^0, or B^1 to match up with each other.

While tensors are not *generally* invariant when changing the order of indices, there are special situations where they are. Tensors that have this special property are called *symmetric* tensors. In symbols, a symmetric tensor has the property

$$T^{\mu\nu} = T^{\nu\mu}.$$

Let me construct one for you. If A and B are 4-vectors, then

$$A^\mu B^\nu + A^\nu B^\mu$$

is a symmetric tensor. If you interchange the indices μ and ν, the value of this expression remains the same. Go ahead and try it. When you rewrite the first term, it becomes the same as the original second term; when you rewrite the second term, it becomes the same as the original first term. The sum of the rewritten terms is the same as the sum of the original terms. If you start with any tensor of rank two, you can construct a symmetric tensor just as we did here.

Symmetric tensors have a special place in general relativity. They're less important in special relativity, but they do come up. For special relativity, the *antisymmetric* tensor is more important. Antisymmetric tensors have the property

$$F^{\mu\nu} = -F^{\nu\mu}.$$

In other words, when you reverse the indices, each component has the same absolute value but changes its sign. To construct an antisymmetric tensor from two 4-vectors, we can write

$$A^{\mu}B^{\nu} - A^{\nu}B^{\mu}.$$

This is almost the same as our trick for constructing a symmetric tensor, except we have a minus sign between the two terms instead of a plus sign. The result is a tensor whose components change sign when you interchange μ and ν. Go ahead and try it.

Antisymmetric tensors have fewer independent components than symmetric tensors. The reason is that their diagonal components must be zero.[6] If $F^{\mu\nu}$ is antisymmetric, each diagonal component must equal its own negative. The only way that can happen is for the diagonal components to equal zero. For example,

$$F^{00} = -F^{00} = 0.$$

The two indices are equal (which is what it means to be a diagonal component), and zero is the only number that equals its own negative. If you think of a rank two tensor as a matrix (a two-dimensional array), then a matrix representing an antisymmetric tensor has zeros all along the diagonal.

---

[6] Diagonal components are components whose two indices are equal to each other. For example, A^{00}, A^{11}, A^{22}, and A^{33} are diagonal components.

6.2.6   An Antisymmetric Tensor

I mentioned earlier that the electric and magnetic fields combine to form a tensor. In fact, they form an antisymmetric tensor. We will eventually derive the tensor nature of the fields $\vec{E}$ and $\vec{B}$, but for now just accept the identification as an illustration.

The names of the elements of the antisymmetric tensor are suggestive, but for now they're just names. The diagonal elements are all zero. We'll call it $F^{\mu\nu}$. It will be convenient to write out the downstairs (covariant) components, $F_{\mu\nu}$:

$$F_{\mu\nu} = \begin{pmatrix} 0 & -E_1 & -E_2 & -E_3 \\ +E_1 & 0 & +B_3 & -B_2 \\ +E_2 & -B_3 & 0 & +B_1 \\ +E_3 & +B_2 & -B_1 & 0 \end{pmatrix}. \tag{6.11}$$

This tensor plays a key role in electromagnetism, where $\vec{E}$ stands for electric field, and $\vec{B}$ stands for magnetic field. We'll see that the electric and magnetic fields combine to form an antisymmetric tensor. The fields $\vec{E}$ and $\vec{B}$ are not independent of each other. In the same way that x can get mixed up with t under a Lorentz transformation, $\vec{E}$ can get mixed up with $\vec{B}$. What you see as a pure electric field, I might see as having some magnetic component. We haven't shown this yet, but we will soon. This is what I meant earlier when I said that electric and magnetic forces transform into one another.

6.3   Electromagnetic Fields

Let's do some physics! We could begin by studying Maxwell's equations, the equations of motion that govern electromagnetic fields. We'll do that in Lecture 8. For now, we'll study the motion of a charged particle *in* an electromagnetic field. The equation that governs this motion is called the Lorentz force law. Later we'll derive this law from an action principle, combined with relativistic ideas. The nonrelativistic (low-velocity) version of the Lorentz force law is

$$m\vec{a} = e\left[\vec{E} + \vec{v} \times \vec{B}\right], \qquad (6.12)$$

where e is the particle's charge and the symbols $\vec{a}$, $\vec{E}$, $\vec{v}$, and $\vec{B}$ are ordinary 3-vectors.[7] The left side is mass times acceleration; the right side must therefore be force. There are two contributions to the force, electric and magnetic. Both terms are proportional to the electric charge e.

In the first term the electric charge of a particle e multiplies the electric field $\vec{E}$. The other term is the magnetic force, in which the charge multiplies the cross product of the particle's velocity $\vec{v}$ with the surrounding magnetic field $\vec{B}$. If you don't know what a cross product is, please take the time to learn. The Appendix contains a brief definition as does the first book in this series, *Classical Mechanics* (Lecture 11). There are many other references as well.

We'll do much of our work with ordinary 3-vectors.[8] Then we'll extend our results to 4-vectors; we will derive the full-blown relativistic version of the Lorentz force law. We're going to find out that the two terms on the right side of Eq. 6.12 are really part of the same term when written in a Lorentz invariant form.

---

[7] Eq. 6.12 is often written with the speed of light shown explicitly. You'll sometimes see it written using $\vec{v}/c$ instead of $\vec{v}$ by itself. Here we'll use units for which the speed of light is 1.

[8] But we'll introduce a 4-vector potential A_μ early in the game.

6.3.1   The Action Integral and the Vector Potential

Back in Lecture 4, I showed you how to construct a Lorentz invariant action (Eq. 4.28) for a particle moving in a scalar field. Let's quickly review the procedure. We began with the action for a free particle in Eq. 4.27,

$$free\ particle\ action = - \int m d\tau,$$

and added a term representing the effect of the field on the particle,

$$new\ term = - \int \phi(t, x) d\tau.$$

This makes a fine theory of a particle interacting with a scalar field, but it's not the theory of a particle in an electromagnetic field. What options do we have for replacing this interaction with something that might yield the Lorentz force law? That is the goal for the rest of this lecture: to derive the Lorentz force law, Eq. 6.12, from an invariant action principle.

How can we modify this Lagrangian to describe the effects of an electromagnetic field? Here there is a bit of a surprise. One might think that the correct procedure is to construct a Lagrangian that involves the particle's coordinate and velocity components, and that depends on the $\vec{E}$ and $\vec{B}$ fields in a manner similar to the action for a particle in the presence of a scalar field. Then, if all goes well, the Euler-Lagrange equation for the motion would involve a force given by the Lorentz force law. Surprisingly, this turns out to be impossible. To make progress we have to introduce another field called the vector potential. In some sense, the electric and magnetic fields are derived quantities constructed from the more basic vector potential, which itself is a 4-vector called $A_\mu(t, x)$. Why a 4-vector? All I can do at this point is ask you to wait a little; you will see that the end justifies the means.

How can we use $A_\mu(t, x)$ to construct an action for a particle in an electromagnetic field? The field $A_\mu(t, x)$ is a 4-vector with a lower index. It seems natural to take a small segment of the trajectory, the 4-vector dX^μ, and combine it with $A_\mu(t, x)$ to make an infinitesimal scalar quantity associated with that segment. For each trajectory segment, we form the quantity

$$dX^\mu A_\mu(t, x).$$

Then we add them up; that is, we integrate from one end to the other, from point a to point b:

$$\int_a^b dX^\mu A_\mu(t, x).$$

Because the expression $dX^\mu A_\mu(t, x)$ is a scalar, every observer agrees about its value on each little segment. If we add quantities that we agree about, we'll continue to agree about them, and we'll get the same answer for the action. To conform to standard conventions, I'll multiply this action by a constant, e:

$$e \int_a^b dX^\mu A_\mu(t, x).$$

I'm sure you've already guessed that e is the electric charge. This is the other term in the action for a particle moving in an electromagnetic field. Let's collect both parts of the action integral:

$$Action = \int_a^b -m\sqrt{1 - \dot{x}^2}\, dt + e \int_a^b dX^\mu A_\mu(t, x). \quad (6.13)$$

6.3.2   The Lagrangian

Both terms in Eq. 6.13 were derived from Lorentz invariant constructions: the first term is proportional to the proper time along the trajectory and the second is constructed from the invariant $dX^\mu A_\mu$. Whatever follows from this action must be Lorentz invariant even if it may not be obvious.

The first term is an old friend and corresponds to the free-particle Lagrangian

$$-m\sqrt{1 - \dot{x}^2}.$$

The second term is new and hopefully will give rise to the Lorentz force law, although at the moment it looks quite unrelated. In its current form, the term

$$e \int_a^b dX^\mu A_\mu(t, x)$$

is not expressed in terms of a Lagrangian, because the integral is not taken over coordinate time dt. That is easily remedied; we just multiply and divide by dt to rewrite it as

$$e \int_a^b \frac{dX^\mu}{dt} A_\mu(t, x)\, dt. \qquad (6.14)$$

In this form, the new term is now the integral of a Lagrangian. Calling the integrand $\mathcal{L}_{int}$, where *int* stands for interaction, the Lagrangian becomes

$$\mathcal{L}_{int} = e \frac{dX^\mu}{dt} A_\mu(t, x). \qquad (6.15)$$

Now let's notice that the quantities

$$\frac{dX^\mu}{dt}$$

come in two different flavors. The first is

$$\frac{dX^0}{dt}.$$

Recalling that X^0 and t are the same thing, we can write

$$\frac{dX^0}{dt} = 1.$$

The other flavor corresponds to the index μ taking one of the values $(1, 2, 3)$, that is, one of the three space directions. In that case we recognize

$$\frac{dX^p}{dt}$$

(with Latin index p) to be a component of the ordinary velocity,

$$\frac{dX^p}{dt} = v_p.$$

If we combine the two types of terms, the interaction Lagrangian becomes

$$\mathcal{L}_{int} = e\dot{X}^p A_p(t, x) + eA_0(t, x), \qquad (6.16)$$

where the repeated Latin index p means that we sum from $p = 1$ to $p = 3$. Because the quantities $\dot{X}^p$ are the components of velocity v_p, the expression $\dot{X}^p A_p(t, x)$ is nothing but the dot product of the velocity with the spatial part of the vector potential. Therefore we could write Eq. 6.16 in a form that may be more familiar,

$$\mathcal{L}_{int} = e\vec{v} \cdot \vec{A} + eA_0.$$

Summarizing all these results, the action integral for a charged particle now looks like this:

$$Action = \int_a^b -m\sqrt{1 - \dot{x}^2}\, dt \; + \; e \int_a^b \left[A_0(t, x) + \dot{X}^p A_p(t, x) \right] dt.$$

$$(6.17)$$

Once again, in more familiar notation, it's

$$Action = \int_a^b -m\sqrt{1 - v^2}\, dt \; + \; e \int_a^b \left(A_0(t, x) + \vec{v} \cdot \vec{A}(t, x) \right) dt.$$

Now that the entire action is expressed as an integral over coordinate time dt, we can easily identify the Lagrangian as

$$\mathcal{L} = -m\sqrt{1 - \dot{x}^2} + eA_0(t, x) + e\dot{X}^n A_n(t, x). \qquad (6.18)$$

Exactly as we did for scalar fields, I'm going to imagine that A_p is a known function of t and x, and that we're just exploring the motion of the particle in that known field. What can we do with Eq. 6.18? Write the Euler-Lagrange equations, of course!

6.3.3   Euler-Lagrange Equations

Here are the Euler-Lagrange equations for particle motion once again, for easy reference:

$$\frac{d}{dt} \frac{\partial \mathcal{L}}{\partial \dot{X}^p} = \frac{\partial \mathcal{L}}{\partial X^p}. \qquad (6.19)$$

Remember that this is shorthand for three equations, one

equation for each value of p.[9] Our goal is to write the Euler-Lagrange equations based on Eq. 6.18 and show that they look like the Lorentz force law. This amounts to substituting $\mathcal{L}$ from Eq. 6.18 into Eq. 6.19. The calculation is a bit longer than for the scalar field case. I'll hold your hand and take you through the steps. Let's start by evaluating

$$\frac{\partial \mathcal{L}}{\partial \dot{X}^p},$$

also known as the *canonical momentum*. There must be a contribution from the first term in Eq. 6.18, because that term contains $\dot{x}$. In fact, we already evaluated this derivative back in Lecture 3. Eq. 3.30 shows the result (with different variable names and in conventional units). In relativistic units, the right side of Eq. 3.30 is equivalent to

$$m\frac{\dot{X}^p}{\sqrt{1 - \dot{x}^2}}.$$

That's the derivative of the *first* term of $\mathcal{L}$ (Eq. 6.18) with respect to the pth component of velocity. The second term of $\mathcal{L}$ does not contain $\dot{x}$ explicitly, so its partial derivative with respect to $\dot{x}$ is zero. However, the third term does contain $\dot{X}^p$, and its partial derivative is

$$eA_p(t, x).$$

Putting these terms together gives us the canonical

---

[9] When an upstairs index appears in the denominator of a derivative, it adds a downstairs index to the result of that derivative. Any time we have a Latin index in an expression, we can move that index upstairs or downstairs as we please. That's because Latin indices represent space components.

momentum,

$$\frac{\partial \mathcal{L}}{\partial \dot{X}^p} = m \frac{\dot{X}_p}{\sqrt{1 - \dot{x}^2}} + eA_p(t, x). \qquad (6.20)$$

We sneaked a small change of notation into Eq. 6.20 by writing $\dot{X}_p$ instead of $\dot{X}^p$. This makes no difference because space components such as p can be moved upstairs or downstairs at will without changing their value. The downstairs index is consistent with the left side of the equation and will be convenient for what follows. Eq. 6.19 now instructs us to take the time derivative, which is

$$\frac{d}{dt}\frac{\partial \mathcal{L}}{\partial \dot{X}^p} = \frac{d}{dt}\left[m \frac{\dot{X}_p}{\sqrt{1 - \dot{x}^2}} + eA_p(t, x) \right].$$

That's the full left side of the Euler-Lagrange equation. What about the right side? The right side is

$$\frac{\partial \mathcal{L}}{\partial X^p}.$$

In what way does $\mathcal{L}$ depend on X_p? The second term, $eA_0(t, x)$, clearly depends on X_p. Its derivative is

$$e\frac{\partial A_0}{\partial X^p}.$$

There's only one more term to account for: $e\dot{X}^p A_p(t, x)$. This term is mixed in that it depends on both velocity and position. But we already accounted for the velocity dependence on the left side of the Euler-Lagrange equation. Now, for the right side, we take the partial derivative with respect to X^p, which is

$$e\dot{X}^n \frac{\partial A_n}{\partial X^p}.$$

Now we can write the full right side of the Euler-Lagrange equation,

$$\frac{\partial \mathcal{L}}{\partial X^p} = e\frac{\partial A_0}{\partial X^p} + e\dot{X}^n\frac{\partial A_n}{\partial X^p}.$$

Setting this equal to the left side gives us the result

$$\frac{d}{dt}\left[m\frac{\dot{X}_p}{\sqrt{1-\dot{x}^2}} + eA_p(t,x) \right] = e\frac{\partial A_0}{\partial X^p} + e\dot{X}^n\frac{\partial A_n}{\partial X^p}. \qquad (6.21)$$

Does this look familiar? The first term on the left side resembles mass times acceleration, except for the square root in the denominator. For low velocities, the square root is very close to 1, and the term literally becomes mass times acceleration. We'll have more to say about the second term, $\frac{d}{dt}eA_p(t,x)$, later on.

You may not recognize the first term on the right side of Eq. 6.21 but it's actually familiar to many of you. However, to see that we have to use different notation. Historically, the time component of the vector potential $A_0(t,x)$ was called $-\phi(t,x)$ and ϕ was called the electrostatic potential. So we can write

$$e\frac{\partial A_0}{\partial X^p} = -e\frac{\partial \phi}{\partial X^p},$$

and the corresponding term in Eq. 6.21 is just minus the electric charge times the gradient of the electrostatic potential. In elementary accounts of electromagnetism, the electric field is minus the gradient of ϕ.

Let's review the structure of Eq. 6.21. The left side has a free (nonsummed) covariant index p. The right side also has a free covariant index p, as well as a summation index n. This is the Euler-Lagrange equation for each component of position, X^1, X^2, and X^3. It doesn't yet look like something we might recognize, but it soon will.

Our next move is to evaluate the time derivatives on the left side. The first term is easy; we'll simply move the m outside the derivative, which gives

$$m \frac{d}{dt} \frac{\dot{X}_p}{\sqrt{1 - \dot{x}^2}}.$$

How do we differentiate the second term, $eA_p(t, x)$, with respect to time? This is a little tricky. Of course, $A_p(t, x)$ may depend explicitly on time. But even if it doesn't, we can't assume that it's constant, because the *position* of the particle is not constant; it changes with time because the particle moves. Even if $A_p(t, x)$ does not depend explicitly on time, its value changes over time as it tracks the motion of the particle. As a result, the derivative generates two terms. The first term is the explicit derivative of A_p with respect to t,

$$e \frac{\partial A_p(t, x)}{\partial t},$$

where the constant e just comes along for the ride. The second term accounts for the fact that $A_p(t, x)$ also depends *implicitly* on time. This term is the change in $A_p(t, x)$ when X^n changes, times $\dot{X}^n$. Putting these terms together results in

$$e \frac{\partial A_p(t, x)}{\partial t} + e \frac{\partial A_p(t, x)}{\partial X^n} \dot{X}^n.$$

Collecting all three terms of the time derivative, the left side of Eq. 6.21 now looks like this:

$$m \frac{d}{dt} \frac{\dot{X}_p}{\sqrt{1 - \dot{x}^2}} + e \frac{\partial A_p(t, x)}{\partial t} + e \frac{\partial A_p(t, x)}{\partial X^n} \dot{X}^n,$$

and we can rewrite the equation as

$$m \frac{d}{dt} \frac{\dot{X}_p}{\sqrt{1 - \dot{x}^2}} + e \frac{\partial A_p}{\partial t} + e \frac{\partial A_p}{\partial X^n} \dot{X}^n$$

$$= e \frac{\partial A_0}{\partial X^p} + e \dot{X}^n \frac{\partial A_n}{\partial X^p}. \tag{6.22}$$

To avoid clutter, I've decided to write A_p instead of $A_p(t, x)$. Just remember that every component of A depends on all four spacetime coordinates. The equation is well formed because each side has a summation index n and a covariant free index p.[10]

This is a lot to assimilate. Let's take a breath and look at what we have so far. The first term in Eq. 6.22 is the relativistic generalization of "mass times acceleration." We'll leave this term on the left side. After moving all the other terms to the right side, here's what we get:

$$m \frac{d}{dt} \frac{\dot{X}_p}{\sqrt{1 - \dot{x}^2}} = -e \frac{\partial A_p}{\partial t} - e \frac{\partial A_p}{\partial X^n} \dot{X}^n + e \frac{\partial A_0}{\partial X^p} + e \dot{X}^n \frac{\partial A_n}{\partial X^p}. \tag{6.23}$$

If the left side is the relativistic generalization of mass times acceleration, the right side can only be the relativistic force on the particle.

There are two kinds of terms on the right side: terms that are proportional to velocity (because they contain $\dot{X}^n$ as a factor), and terms that are not. Grouping similar terms together on the right side results in

$$m \frac{d}{dt} \frac{\dot{X}_p}{\sqrt{1 - \dot{x}^2}} = e \left(\frac{\partial A_0}{\partial X^p} - \frac{\partial A_p}{\partial t} \right) + e \dot{X}^n \left(\frac{\partial A_n}{\partial X^p} - \frac{\partial A_p}{\partial X^n} \right). \tag{6.24}$$

---

[10] It would be okay if each side of the equation had a different summation index.

Art: *Ouch, my head hurts. I don't recognize any of this. I thought you said we were going to get the Lorentz force law. You know,*

$$\vec{F} = e\vec{E} + e\vec{v} \times \vec{B}.$$

All I see in common with Eq. 6.24 is the electric charge e.

Lenny: *Hang on, Art, we're getting there.*

Look at the left side of Eq. 6.24. Suppose the velocity is small so that we can ignore the square root in the denominator. Then the left side is just mass times acceleration, which Newton would call the force. So we have our $\vec{F}$.

On the right side we have two terms, one of which contains a velocity $\dot{X}^n$, which I could call v_n. The other term doesn't depend on the velocity.

Art: *I think I see the light, Lenny! Is the term without the velocity $e\vec{E}$?*

Lenny: *Now you're cooking, Art. Go for it; what's the term with the velocity?*

Art: *Holy cow! It must be ... Yes! It must be $e\vec{v} \times \vec{B}$.*

Art's right. Consider the term that doesn't depend on velocity, namely

$$e\left(\frac{\partial A_0}{\partial X^p} - \frac{\partial A_p}{\partial t}\right).$$

If we define the electric field component E_p to be

$$E_p = \left(\frac{\partial A_0}{\partial X^p} - \frac{\partial A_p}{\partial t}\right), \tag{6.25}$$

we get the first term of the Lorentz law, $e\vec{E}$.

The velocity-dependent term is a bit harder unless you are a master of cross products, in which case you can see that it's $e\vec{v} \times \vec{B}$ in component form. In case you're not a master of cross products, we'll work it out next.[11]

Let's first consider the z component of $\vec{v} \times \vec{B}$,

$$(\vec{v} \times \vec{B})_z = v_x B_y - v_y B_x. \tag{6.26}$$

We want to compare this with the z component of

$$\dot{X}^n \left(\frac{\partial A_n}{\partial X^p} - \frac{\partial A_p}{\partial X^n} \right), \tag{6.27}$$

which we get by identifying the index p with the direction z. The index n is a summation index, which means we need to substitute the values $(1, 2, 3)$, or equivalently (x, y, z), and then add up the results. There's no shortcut; just plug in these values and you'll get

$$v_x \left(\frac{\partial A_x}{\partial z} - \frac{\partial A_z}{\partial x} \right) + v_y \left(\frac{\partial A_y}{\partial z} - \frac{\partial A_z}{\partial y} \right). \tag{6.28}$$

Exercise 6.2: Expression 6.28 was derived by identifying the index p with the z component of space, and then summing over n for the values $(1, 2, 3)$. Why doesn't Expression 6.28 contain a v_z term?

Now all we have to do is to match Expression 6.28 with the right side of Eq. 6.26. We can do this by making the identifications

$$B_y = \frac{\partial A_x}{\partial z} - \frac{\partial A_z}{\partial x}$$

---

[11] The summary of 3-vector operators in Appendix B should be helpful for navigating the rest of this section.

and

$$B_x = -\left(\frac{\partial A_y}{\partial z} - \frac{\partial A_z}{\partial y}\right).$$

The minus sign in the second equation is due to the minus sign in the second term of Eq. 6.26. We can do exactly the same thing for the other components, and we find that

$$B_x = \frac{\partial A_z}{\partial y} - \frac{\partial A_y}{\partial z}$$

$$B_y = \frac{\partial A_x}{\partial z} - \frac{\partial A_z}{\partial x}$$

$$B_z = \frac{\partial A_y}{\partial x} - \frac{\partial A_x}{\partial y}. \tag{6.29}$$

There is a shorthand notation for these equations that you should be familiar with.[12] Eqs. 6.29 are summarized by saying that $\vec{B}$ is the curl of $\vec{A}$,

$$\vec{B} = \vec{\nabla} \times \vec{A}. \tag{6.30}$$

How should one think about Eqs. 6.25 and 6.30? One way is to say that the fundamental quantity defining an electromagnetic field is the vector potential, and that Eqs. 6.25 and 6.30 define derived objects that we recognize as electric and magnetic fields. But one might respond that the quantities that are most physical, the ones directly measured in the laboratory, are E and B, and the vector potential is just a trick for describing them. Either way is fine: it's a Monday-Wednesday-Friday versus Tuesday-Thursday-Saturday kind of thing. But whatever your philosophy about the primacy of

---

[12] It's described in Appendix B.

A versus (E, B) happens to be, the experimental fact is that charged particles move according to the Lorentz force law. In fact, we can write Eq. 6.24 in the fully relativistically invariant form,

$$m\frac{d}{dt}\frac{\dot{X}_p}{\sqrt{1-\dot{x}^2}} = e(\vec{E} + \vec{v} \times \vec{B})_p. \qquad (6.31)$$

6.3.4 Lorentz Invariant Equations

Eq. 6.31 is the Lorentz invariant form of the equation of motion for a charged particle; the left side is a relativistic form of mass times acceleration, and the right side is the Lorentz force law. We know this equation is Lorentz invariant because of the way we defined the Lagrangian. However, it is not *manifestly* invariant; it is not obviously invariant based on the structure of the equation. When a physicist says that an equation is manifestly invariant under Lorentz transformations, she means that it is written in a four-dimensional form with upstairs and downstairs indices matching up in the right way. In other words, it is written as a tensor equation with both sides transforming the same way. That's our goal for the rest of this lecture.

Let's go all the way back to the previous form of the Euler-Lagrange equations (Eq. 6.24), which I'll rewrite here for convenience.

$$m\frac{d}{dt}\frac{\dot{X}_p}{\sqrt{1-\dot{x}^2}} = e\left(\frac{\partial A_0}{\partial X^p} - \frac{\partial A_p}{\partial t}\right) + e\dot{X}^n\left(\frac{\partial A_n}{\partial X^p} - \frac{\partial A_p}{\partial X^n}\right).$$

Back in Lectures 2 and 3 (see Eqs. 2.16 and 3.9), we found that the quantities

$$\frac{\dot{X}^p}{\sqrt{1-\dot{x}^2}}$$

are the space components of the velocity 4-vector. In other words,

$$\frac{\dot{X}^p}{\sqrt{1-\dot{x}^2}} = \frac{dX^p}{d\tau}.$$

Because p is a Latin index, we can also write

$$\frac{\dot{X}_p}{\sqrt{1-\dot{x}^2}} = \frac{dX_p}{d\tau}. \tag{6.32}$$

Therefore (ignoring the factor m for now), the left side of Eq. 6.24 can be written as

$$\frac{d}{dt}\left(\frac{dX^p}{d\tau}\right).$$

These are not components of a 4-vector. However, they can be put into 4-vector form if we multiply them by $dt/d\tau$,

$$\frac{dt}{d\tau} \cdot \frac{d}{dt}\frac{dX^p}{d\tau} = \frac{d^2 X^p}{d\tau^2}.$$

These are the *space* components of a 4-vector—the relativistic acceleration. The full 4-vector

$$\frac{d^2 X^\mu}{d\tau^2}$$

has four components. Let's therefore multiply both sides of Eq. 6.24 by $dt/d\tau$. We'll replace t with X^0, and we'll write $dX^0/d\tau$ instead of $dt/d\tau$. The result is

$$m\frac{d^2 X_p}{d\tau^2} = e\frac{dX^0}{d\tau}\left(\frac{\partial A_0}{\partial X^p} - \frac{\partial A_p}{\partial X^0}\right) + e\frac{dX^n}{d\tau}\left(\frac{\partial A_n}{\partial X^p} - \frac{\partial A_p}{\partial X^n}\right).$$

Recalling our conventions for index ranges, we could try to rewrite the right side of this equation as a single term by replacing the Latin indices with Greek indices that range from 0 through 4:

$$m\frac{d^2 X_p}{d\tau^2} = e\frac{dX^\nu}{d\tau}\left(\frac{\partial A_\nu}{\partial X^\mu} - \frac{\partial A_\mu}{\partial X^\nu}\right).$$

However, this would create a problem. Each side of the equation would have a downstairs free index, which is fine. But the left side would have a Latin free index p, while the right side would have a Greek free index μ. Indices can be renamed, of course, but we can't match up a Latin free index on one side with a Greek free index on the other because they have different ranges. To write this equation properly, we need to replace X_p with X_μ on the left side. The properly formed equation is

$$m\frac{d^2 X_\mu}{d\tau^2} = e\frac{dX^\nu}{d\tau}\left(\frac{\partial A_\nu}{\partial X^\mu} - \frac{\partial A_\mu}{\partial X^\nu}\right). \qquad (6.33)$$

If we set the index μ to be the space index p, then Eq. 6.33 just reproduces Eq. 6.24. In fact, Eq. 6.33 is the *manifestly invariant* equation of motion for a charged particle. It's manifestly invariant because all the objects in the equations are 4-vectors and all the repeated indices are properly contracted.

There is one important subtlety: We started with Eq. 6.24, which represents only three equations for the space components, labeled by the index p. When we allow the index μ to range over all four directions of Minkowski space, Eq. 6.33 gives one extra equation for the time component.

But is this new equation (the zeroth equation) correct? By making sure at the beginning that the action was a scalar, we guaranteed that our equations of motion would be Lorentz invariant. If the equations are Lorentz invariant, and the three space components of a certain 4-vector are equal to the three space components of some other 4-vector, then we know *automatically* that their zeroth components (the time components) must also match. Because we've built a Lorentz invariant theory where the three space equations are correct, the time equation must also be correct.

This logic pervades everything in modern physics: Make sure your Lagrangian respects the symmetries of the problem. If the symmetry of the problem is Lorentz symmetry, make sure your Lagrangian is Lorentz invariant. This guarantees that the equations of motion will also be Lorentz invariant.

6.3.5   Equations with 4-Velocity

Let's massage Eq. 6.33 into a slightly different form that will come in handy later on. Recall that the 4-vector U^μ, defined as

$$U^\mu = \frac{dX^\mu}{d\tau},$$

is called the 4-velocity. Making this substitution in Eq. 6.33 (with a lowered μ index on the left side) results in

$$m\frac{dU_\mu}{d\tau} = e\left(\frac{\partial A_\nu}{\partial X^\mu} - \frac{\partial A_\mu}{\partial X^\nu}\right)U^\nu. \qquad (6.34)$$

The derivative

$$\frac{dU_\mu}{d\tau}$$

that appears on the left side is called the 4-acceleration. The spatial part of this equation in terms of U is

$$m\frac{dU_p}{d\tau} = e\left(\frac{\partial A_0}{\partial X^p} - \frac{\partial A_p}{\partial X^0}\right)U^0 + e\left(\frac{\partial A_n}{\partial X^p} - \frac{\partial A_p}{\partial X^n}\right)U^n. \quad (6.35)$$

6.3.6 Relationship of A_μ to $\vec{E}$ and $\vec{B}$

Let's summarize the relation between the vector potential A_μ and the familiar electric and magnetic fields $\vec{E}$ and $\vec{B}$.

Historically, $\vec{E}$ and $\vec{B}$ were discovered through experiments on charges and electric currents (electric currents are just charges in motion). Those discoveries culminated in Lorentz's synthesis of the Lorentz force law. But try as you may, when formulated in terms of $\vec{E}$ and $\vec{B}$ there is no way to express the dynamics of charged particles as an action principle. For that, the vector potential is essential. Writing the Euler-Lagrange equations and comparing them with the Lorentz force law provided the following equations relating the $\vec{E}$ and $\vec{B}$ fields to A_μ:

$$E_x = \frac{\partial A_0}{\partial x} - \frac{\partial A_x}{\partial t}$$

$$E_y = \frac{\partial A_0}{\partial y} - \frac{\partial A_y}{\partial t}$$

$$E_z = \frac{\partial A_0}{\partial z} - \frac{\partial A_z}{\partial t}$$

$$B_x = \frac{\partial A_z}{\partial y} - \frac{\partial A_y}{\partial z}$$

$$B_y = \frac{\partial A_x}{\partial z} - \frac{\partial A_z}{\partial x}$$

$$B_z = \frac{\partial A_y}{\partial x} - \frac{\partial A_x}{\partial y}. \tag{6.36}$$

These equations have a certain symmetry that begs us to put them into tensor form. We'll do that in the next lecture.

6.3.7   The Meaning of U^μ

At the end of this lecture a student raised his hand and remarked that we had done a good deal of heavy mathematical lifting to get to Eq. 6.34. But could we come back to earth and discuss the meaning of the relativistic 4-velocity U^μ? In particular, what about the zeroth equation—the equation for the time component of the relativistic acceleration?

Sure, let's do it. First the space components, shown in Eq. 6.35. The left side is just a relativistic generalization of mass times acceleration—the ma of Newton's $F = ma$. If the particle is moving slowly, then it's exactly that. But the equation also works if the particle is moving fast. The right side is the relativistic version of the Lorentz force; the first term is the electric force, and the second term is the magnetic force that acts on a moving charge. The zeroth equation is the one that needs explanation. Let's write it as

$$m \frac{dU_0}{d\tau} = e\left(\frac{\partial A_n}{\partial X^0} - \frac{\partial A_0}{\partial X^n}\right)\frac{dX^n}{d\tau}. \qquad (6.37)$$

Notice that we omitted the first term on the right side (where $\mu = 0$) because

$$\left(\frac{\partial A_0}{\partial X^0} - \frac{\partial A_0}{\partial X^0}\right) = 0.$$

Let's start with the left side of Eq. 6.37. It's something a little unfamiliar: the time component of the relativistic acceleration. To interpret it we recall (Eqs. 3.36 and 3.37) that the relativistic kinetic energy is just

$$mU^0 = \frac{m}{\sqrt{1 - v^2}}.$$

Using conventional units to make the speed of light explicit, this becomes

$$mU^0 = \frac{mc^2}{\sqrt{1 - v^2/c^2}}.$$

Just to keep the bookkeeping straight, mU_0 is *minus* the kinetic energy because lowering the time index changes the sign. Therefore, apart from the minus sign, the left side of Eq. 6.37 is the time rate of change of the kinetic energy. The rate of change of kinetic energy is the work done on the particle per unit time by whatever force is acting on it.

Next, we consider the right side of Eq. 6.37; it has the form

$$-e\vec{E} \cdot \vec{v}.$$

Using the fact that $e\vec{E}$ is the electric force on the particle, we can express the right side of the equation as

$$-\vec{F} \cdot \vec{v}.$$

From elementary mechanics the dot product of the force and the velocity is exactly the work (per unit time) done by force on a moving particle. So we see that the zeroth equation is just the energy balance equation, expressing the fact that the change of kinetic energy is the work done on a system.

What happened to the magnetic force? How come we did not have to include it in calculating the work? One answer that you probably learned in high school is this:

Magnetic forces do no work.

But there is a simple explanation: The magnetic force is always perpendicular to the velocity, so it does not contribute to $\vec{F} \cdot \vec{v}$.

6.4   Interlude on the Field Tensor

The first question a modern physicist might ask about a new quantity is, how does it transform? In particular, how does it transform under Lorentz transformations? The usual answer is that the new quantity—whatever it refers to—is a tensor with so many upstairs indices and so many downstairs indices. The electric and magnetic fields are no exception.

Art: *Lenny, I have a big problem. The electric field has three space components and no time component. How the devil can it be any kind of four-dimensional tensor? Same problem with the magnetic field.*

Lenny: *You're forgetting, Art:* $3 + 3 = 6$.

Art: *You mean there's some kind of tensor that has six components? I don't get it; a 4-vector has four components, a tensor with two indices has* $4 \times 4 = 16$ *components. What's the trick?*

The answer to Art's question is that while it's true that tensors with two indices have sixteen components, antisymmetric tensors only have six *independent* components. Let's just review the idea of an antisymmetric tensor.

Suppose we have two 4-vectors C and B. There are two ways we can put them together to make a two-indexed tensor: $C_\mu B_\nu$ and $B_\mu C_\nu$. By adding and subtracting these two, we can build a symmetric and an antisymmetric tensor,

$$S_{\mu\nu} = C_\mu B_\nu + B_\mu C_\nu$$

and

$$A_{\mu\nu} = C_\mu B_\nu - B_\mu C_\nu.$$

In particular, let's count the number of independent components of an antisymmetric tensor. First of all, the diagonal elements all vanish. That means of the sixteen components, only twelve survive.

But there are additional constraints; the off-diagonal elements come in equal and opposite pairs. For example, $A_{02} = -A_{20}$. Only half of the twelve are independent, so that leaves six. Yes Art, there are tensors that have only six independent components.

Now look at Eq. 6.36. All of the right sides have the form

$$F_{\mu\nu} = \frac{\partial A_\nu}{\partial X^\mu} - \frac{\partial A_\mu}{\partial X^\nu}, \tag{6.38}$$

or, equivalently,

$$F_{\mu\nu} = \partial_\mu A_\nu - \partial_\nu A_\mu. \tag{6.39}$$

$F_{\mu\nu}$ is obviously a tensor and what is more, it is antisymmetric. It has six independent components and they are precisely the components of $\vec{E}$ and $\vec{B}$. Comparing with Eq. 6.36, we can list the positive components:

$$F_{10} = E_x \qquad\qquad F_{12} = B_z$$

$$F_{20} = E_y \qquad\qquad F_{23} = B_x$$

$$F_{30} = E_z \qquad\qquad F_{31} = B_y. \tag{6.40}$$

You can use the antisymmetry to fill in the other components. Now you can see where Eq. 6.11 comes from. I'll write it again here along with all its upstairs-downstairs versions.

$$F_{\mu\nu} = \begin{pmatrix} 0 & -E_x & -E_y & -E_z \\ +E_x & 0 & +B_z & -B_y \\ +E_y & -B_z & 0 & +B_x \\ +E_z & +B_y & -B_x & 0 \end{pmatrix} \qquad (6.41)$$

$$F^{\mu\nu} = \begin{pmatrix} 0 & +E_x & +E_y & +E_z \\ -E_x & 0 & +B_z & -B_y \\ -E_y & -B_z & 0 & +B_x \\ -E_z & +B_y & -B_x & 0 \end{pmatrix} \qquad (6.42)$$

$$F^{\mu}_{\ \nu} = \begin{pmatrix} 0 & +E_x & +E_y & +E_z \\ +E_x & 0 & +B_z & -B_y \\ +E_y & -B_z & 0 & +B_x \\ +E_z & +B_y & -B_x & 0 \end{pmatrix} \qquad (6.43)$$

$$F_{\mu}^{\ \nu} = \begin{pmatrix} 0 & -E_x & -E_y & -E_z \\ -E_x & 0 & +B_z & -B_y \\ -E_y & -B_z & 0 & +B_x \\ -E_z & +B_y & -B_x & 0 \end{pmatrix} \qquad (6.44)$$

You can think of the rows of these matrices being labeled by index μ, with its values $(0, 1, 2, 3)$ running along the left edge from top to bottom. Likewise, the columns are labeled by index ν with the same (unwritten) values running across the top.

Look at the top row of $F_{\mu\nu}$. We can think of these elements as $F_{0\nu}$, because $\mu = 0$ for each of them, while ν takes on its full range of values. These are electric field components. We think of them as "mixed" time and space components because one of their indices is a time index while the other is a space index. The same is true for the leftmost column, and for the same reasons.

Now look at the 3×3 submatrix that *excludes* the topmost row and leftmost column. This submatrix is populated with magnetic field components.

What did we accomplish by organizing the electric and

magnetic fields into a single tensor? A lot. We now know how to answer the following question: Suppose Lenny, in his frame on the moving train, sees a certain electric and magnetic field. What electric and magnetic field does Art see in the frame of the rail station? A charge at rest in the train would have a familiar electric field (Coulomb field) in Lenny's frame. From Art's point of view that same charge is moving, and to compute the field that Art sees is now just a matter of Lorentz-transforming the field tensor. We'll explore this idea further in Section 8.1.1.

Lecture 7

Fundamental Principles and Gauge Invariance

Art: *Why are physicists so hung up on fundamental principles like least action, locality, Lorentz invariance, and—what's the other one?*

Lenny: *Gauge invariance. The principles help us evaluate new theories. A theory that violates them is probably flawed. But sometimes we're forced to rethink what we mean by fundamental.*

Art: *Okay, the speeding railroad car: It's Lorentz invariant. It's racing down the track, so it has all the action it needs (but no more). It stops at every station, so it's local.*

Lenny: *Ugh.*

Art: *And it's gauge invariant; otherwise it falls off the tracks!*

Lenny: *I think* you *just fell off the tracks. Let's slow things down a little.*

Suppose a theoretical physicist wants to construct a theory to explain some newly discovered phenomenon. The new theory is expected to follow certain rules or fundamental principles. There are four principles that seem to govern all physical laws. Briefly, they are:

- The action principle
- Locality
- Lorentz invariance
- Gauge invariance

These principles pervade all of physics. Every known theory, whether it's general relativity, quantum electrodynamics, the standard model of particle physics, or Yang-Mills theory, conforms to them. The first three principles should be familiar, but gauge invariance is new; we haven't seen it before. The main goal of this lecture is to introduce this new idea. We'll start with a summary of all four fundamental principles.

7.1   Summary of Principles

Action Principle

The first rule is that physical phenomena are described by an action principle. We don't know of any exceptions to this pattern. The concept of energy conservation, to cite just one example, is derived entirely through the action principle. The same is true for momentum conservation and for the relation between conservation laws and symmetries in general. If you just write equations of motion, they may make perfectly good sense. But if they're not derived from an action principle, we would give up the guarantee of energy and momentum conservation. Energy conservation, in particular, is a consequence of the action principle, together with the assumption that things are invariant under shifting the time by a fixed amount—a transformation that we call a *time translation*.

So that's our first principle: Look for an action such that the resulting equations of motion describe the phenomena that are discovered in the laboratory. We have seen two kinds of action. The action for particle motion is

$$Action_{particle} = \int dt \mathcal{L}(X, \dot{X}),$$

where $\mathcal{L}$ represents the Lagrangian. The action for field theories is

$$Action_{field} = \int d^4 x \mathcal{L}(\phi, \phi_\mu).$$

For field theories, $\mathcal{L}$ represents the *Lagrangian density*. The word *density* indicates a quantity that's integrated over space as well as time. We've seen how the Euler-Lagrange equations govern both of these cases.

Locality

Locality means that things happening at one place only directly affect conditions nearby in space and time. If you poke a system at some point in time and space, the only *direct* effect is on things in its immediate vicinity. For example, if you strike a violin string at its end point, only its nearest neighbor would feel the effect immediately. Of course, the neighbor's motion would affect *its* neighbor, and so forth down the chain. In time, the effect would be felt along the entire length of the string. But the short-time effect is local.

How do we guarantee that a theory respects locality? Once again, it happens through the action. For example, suppose we're talking about a particle. In that case, the action is an integral over time (dt) along the particle trajectory. To guarantee locality, the integrand—the Lagrangian $\mathcal{L}$—must depend only on the coordinates of the system. For a particle, that means the position components of the particle, and their first time derivatives. Neighboring time points come into play through the time derivatives. After all, derivatives are things that capture relationships between near neighbors. However, higher derivatives are ruled out because they're "less local" than first derivatives.[1]

Field theories describe a field contained in a volume of space and time (Figs. 4.2 and 5.1). The action is an integral not only over time, but also over space (d^4x). In this case, locality says that the Lagrangian depends on the field ϕ and on its partial derivatives with respect to X^μ. We can call these derivatives ϕ_μ. This is enough to guarantee that things only affect their nearby neighbors directly.

You could imagine a world in which poking something in one place has an instantaneous effect in some other place. In that case the Lagrangian would not depend only on nearest neighbors through derivatives but on more complicated things that allow "action at a distance." Locality forbids this.

A quick word about quantum mechanics: Quantum mechanics is outside the scope of this book. Nevertheless, many readers may wonder how—or whether—the locality principle applies to quantum mechanics. Let's be as clear as possible: *It does*.

Quantum mechanical entanglement is often referred to misleadingly as nonlocality. But entanglement is not the same as nonlocality. In our previous book, *Quantum Mechanics*, we explain this in great detail. Entanglement does not mean you can send signals instantaneously from one place to another. Locality is fundamental.

---

[1] They're also ruled out by a large body of theoretical and experimental results.

Lorentz Invariance

A theory must be Lorentz invariant. In other words, the equations of motion should be the same in every reference frame. We've already seen how this works. If we make sure the Lagrangian is a scalar, we guarantee that the theory is Lorentz invariant. In symbols,

$$\mathcal{L} = Scalar.$$

Lorentz invariance includes the idea of invariance under rotations of space.[2]

Gauge Invariance

The last rule is somewhat mysterious and takes some time to fully understand. Briefly, gauge invariance has to do with changes that you can make to the vector potential without affecting the physics. We'll spend the rest of this lecture introducing it.

7.2 Gauge Invariance

An invariance, also called a *symmetry*, is a change in a system that doesn't affect the action or the equations of motion. Let's look at some familiar examples.

7.2.1 Symmetry Examples

The equation $F = m\vec{a}$ is perhaps the best-known equation of motion. It has exactly the same form if you translate the

---

[2] General relativity (not covered in this book) requires invariance under arbitrary coordinate transformations. Lorentz transformations are a special case. Even so, the invariance principle is similar; instead of requiring $\mathcal{L}$ to be a scalar, general relativity requires it to be a scalar density.

origin of coordinates from one point to another. The same is true for rotations of the coordinates. This law is invariant under translations and rotations.

As another example, consider our basic field theory from Lecture 4. The Lagrangian for this theory (Eq. 4.7) was

$$\mathcal{L} = \frac{1}{2}\left[\left(\frac{\partial\phi}{\partial t}\right)^2 - \left(\frac{\partial\phi}{\partial x}\right)^2 - \left(\frac{\partial\phi}{\partial y}\right)^2 - \left(\frac{\partial\phi}{\partial z}\right)^2\right] - V(\phi),$$

which we also expressed as

$$-\frac{1}{2}\left[\partial_\mu\phi\partial^\mu\phi\right] - V(\phi).$$

For now, we'll consider a simplified version, with $V(\phi)$ set to zero and all the space coordinates folded into the single variable x:

$$\mathcal{L} = \frac{1}{2}\left[\left(\frac{\partial\phi}{\partial t}\right)^2 - \left(\frac{\partial\phi}{\partial x}\right)^2\right].$$

The equation of motion (Eq. 4.10) we derived from this Lagrangian was (slightly simplified here)

$$\frac{\partial^2\phi}{\partial t^2} - \frac{\partial^2\phi}{\partial x^2} = 0. \tag{7.1}$$

I've ignored the factor $\frac{1}{c^2}$ in the first term because it's not important for this example. This equation has a number of invariances, including Lorentz invariance. To discover a new invariance, we try to find things about the equation that we can change without changing its content or meaning. Suppose we add a constant to the underlying field:

$$\phi \longrightarrow \phi + c.$$

In other words, suppose we take a field ϕ *that is already a solution* to the equation of motion and just add a constant.

Does the result still satisfy the equation of motion? Of course it does, because the derivatives of a constant are zero. If we know that ϕ satisfies the equation of motion, then $(\phi + c)$ also satisfies it:

$$\frac{\partial^2 (\phi + c)}{\partial t^2} - \frac{\partial^2 (\phi + c)}{\partial x^2} = 0.$$

You can also see this in the Lagrangian,

$$-\partial_\mu \phi \partial^\mu \phi$$

where once again I've ignored the factor $\frac{1}{2}$ for simplicity. What happens to this Lagrangian (and consequently to the action) if we add a constant to ϕ? Nothing! The derivative of a constant is zero. If we have a particular action and a field configuration that minimizes the action, adding a constant to the field makes no difference; it will still minimize the action. In other words, adding a constant to such a field is a symmetry, or an invariance. It's a somewhat different kind of invariance than we've seen before, but it's an invariance all the same.

Now let's recall the slightly more complicated version of this theory, where the term $V(\phi)$ is *not* zero. Back in Lecture 4, we considered the case

$$V(\phi) = \frac{\mu^2}{2} \phi^2,$$

whose derivative with respect to ϕ is

$$\frac{\partial V(\phi)}{\partial \phi} = \mu^2 \phi.$$

With these changes, the Lagrangian becomes

$$\mathcal{L} = \frac{1}{2} \left[\left(\frac{\partial \phi}{\partial t} \right)^2 - \left(\frac{\partial \phi}{\partial x} \right)^2 \right] - \frac{\mu^2}{2} \phi^2, \qquad (7.2)$$

and the equation of motion becomes

$$\frac{\partial^2 \phi}{\partial t^2} - \frac{\partial^2 \phi}{\partial x^2} + \mu^2 \phi = 0. \tag{7.3}$$

What happens to Eq. 7.3 if we add a constant to ϕ? If ϕ is a solution, is $(\phi + c)$ still a solution? It is not. There's no change in the first two terms. But adding a constant to ϕ clearly changes the third term. What about the Lagrangian, Eq. 7.2? Adding a constant to ϕ has no impact on the terms inside the square brackets. But it does affect the rightmost term; ϕ^2 is not the same as $(\phi + c)^2$. With the extra term

$$-\frac{\mu^2}{2}\phi^2$$

in the Lagrangian, adding a constant to ϕ is *not* an invariance.

7.2.2   A New Kind of Invariance

Let's return to the action integral

$$e \int_a^b A_\mu dX^\mu,$$

which we introduced in Lecture 6. Suppose we modify A_μ by adding the four-dimensional gradient of a scalar S. In symbols,

$$A_\mu \longrightarrow A_\mu + \frac{\partial S}{\partial X^\mu}.$$

The preceding sum makes sense because both terms are covariant in the index μ. But does this replacement change the equations of motion? Does it change the orbits that the particle will follow? Does it make any change whatever in the dynamics of the particle? The replacement changes the

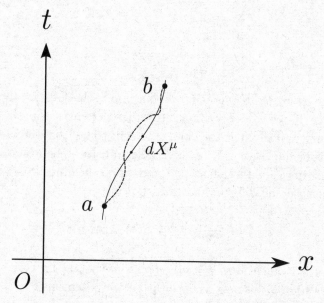

Figure 7.1: Spacetime Trajectory. The solid line is the stationary-action path. The dashed line is the varied path with the same fixed end points.

action in a straightforward way;

$$Action_{original} = e \int_a^b A_\mu dX^\mu \qquad (7.4)$$

becomes

$$Action_{modified} = e \int_a^b A_\mu dX^\mu + e \int_a^b \frac{\partial S}{\partial X^\mu} dX^\mu. \qquad (7.5)$$

What does the new integral—the rightmost integral in Eq. 7.5—represent? In Fig. 7.1, the particle trajectory is broken into little segments, as usual. Consider the change in S along one of these little segments. Straightforward calculus tells us that

$$\frac{\partial S}{\partial X^\mu} \, dX^\mu$$

is the change in S in going from one end of the segment to the other. If we add (or integrate) these changes for all the segments, it gives us the change in S in going from the initial point of the trajectory to the final point. In other words, it's equal to S evaluated at the end point b minus S at the starting point a. In symbols,

$$\int_a^b \frac{\partial S}{\partial X^\mu} \, dX^\mu = S(b) - S(a).$$

S itself is an *arbitrary scalar function* that I just threw in. Does adding this new term to the vector potential change the dynamics of the particle?[3]

We've changed the action in a way that only depends on the end points of the trajectory. But the action principle tells us to search for the minimum action by wiggling the trajectory *subject to the constraint that the end points remain fixed*. Because the end points remain fixed, the $S(b) - S(a)$ term is the same for every trajectory, including the trajectory with stationary action. Therefore, if you find a trajectory with stationary action, it will remain stationary when you make this change to the vector potential. Adding the four-dimensional gradient of a scalar has no effect on the motion of the particle because it only affects the action at the end points. In fact, adding any derivative to the action typically doesn't change anything. We don't even care what S is; our reasoning applies to any scalar function S.

This is the concept of gauge invariance. The vector potential can be changed, in a certain way, without any

_____
[3] Hint: The answer is *No!*

effect on the behavior of the charged particle. Adding $\dfrac{\partial S}{\partial X^\mu}$ to a vector potential is called a *gauge transformation*. The word *gauge* is a historical artifact that has little to do with the concept.

7.2.3   Equations of Motion

Gauge invariance makes a bold claim:

> *Dream up any scalar function you like, add its gradient to the vector potential, and the equations of motion stay exactly the same.*

Can we be sure about that? To find out, just go back to the equations of motion: the Lorentz force law of Eq. 6.33,

$$m\frac{d^2 X_\mu}{d\tau^2} = e\frac{dX^\nu}{d\tau}\left(\frac{\partial A_\nu}{\partial X^\mu} - \frac{\partial A_\mu}{\partial X^\nu}\right).$$

We can rewrite this as

$$m\frac{d^2 X_\mu}{d\tau^2} = eF_{\mu\nu}U^\nu,$$

where

$$F_{\mu\nu} = \frac{\partial A_\nu}{\partial X^\mu} - \frac{\partial A_\mu}{\partial X^\nu} \tag{7.6}$$

and the 4-velocity U^ν is

$$U^\nu = \frac{dX^\nu}{d\tau}.$$

The equations of motion do not directly involve the vector potential. They involve the field tensor $F_{\mu\nu}$, whose elements are the electric and magnetic field components. Any change to the vector potential that does not affect $F_{\mu\nu}$ will not affect the motion of the particle. Our task is to verify that the field tensor $F_{\mu\nu}$ is gauge invariant. Let's see what happens when we add the gradient of a scalar to the vector potentials of Eq. 7.6:

$$F_{\mu\nu} = \frac{\partial\left(A_\nu + \dfrac{\partial S}{\partial X^\nu}\right)}{\partial X^\mu} - \frac{\partial\left(A_\mu + \dfrac{\partial S}{\partial X^\mu}\right)}{\partial X^\nu}. \qquad (7.7)$$

This looks complicated, but it simplifies quite easily. The derivative of a sum is the sum of the derivatives, so we can write

$$F_{\mu\nu} = \frac{\partial A_\nu}{\partial X^\mu} + \frac{\partial\left(\dfrac{\partial S}{\partial X^\nu}\right)}{\partial X^\mu} - \frac{\partial A_\mu}{\partial X^\nu} - \frac{\partial\left(\dfrac{\partial S}{\partial X^\mu}\right)}{\partial X^\nu},$$

or

$$F_{\mu\nu} = \frac{\partial A_\nu}{\partial X^\mu} + \frac{\partial^2 S}{\partial X^\mu \partial X^\nu} - \frac{\partial A_\mu}{\partial X^\nu} - \frac{\partial^2 S}{\partial X^\nu \partial X^\mu}. \qquad (7.8)$$

But we know that the order in which partial derivatives are taken doesn't matter. In other words,

$$\frac{\partial^2 S}{\partial X^\nu \partial X^\mu} = \frac{\partial^2 S}{\partial X^\mu \partial X^\nu}.$$

Therefore the two second-order derivatives in Eq. 7.8 are equal and cancel each other out.[4] The result is

$$F_{\mu\nu} = \frac{\partial A_\nu}{\partial X^\mu} - \frac{\partial A_\mu}{\partial X^\nu}, \qquad (7.9)$$

which is exactly the same as Eq. 7.6. This closes the circle.

---

[4] This property of second partial derivatives is true for the functions that interest us. There do exist functions that do not have this property.

Adding the gradient of a scalar to the 4-vector potential has no impact on the action or on the equations of motion.

7.2.4 Perspective

If our goal was to write equations of motion for particles and electromagnetic fields, why did we bother adding a vector potential, especially if we can change it without affecting the electric and magnetic fields?

The answer is that there's no way to write an action principle for the motion of a particle that does not involve the vector potential. Yet the value of the vector potential at a given point is not physically meaningful and it cannot be measured; if you change it by adding the gradient of a scalar, the physics doesn't change.

Some invariances have obvious physical meaning. It's not hard to imagine two reference frames in the same problem and translating between them: two physical reference frames, yours and mine. Gauge invariance is different. It's not about transformations of coordinates. It's a *redundancy of the description*. Gauge invariance means there are many descriptions, all of which are equivalent to each other. What's new is that it involves a function of position. For example, when we rotate coordinates, we don't rotate differently at different locations. That sort of rotation would not define an invariance in ordinary physics. We would rotate once and only once, by some specific angle, in a way that does *not* involve a function of position. On the other hand, a gauge transformation involves a whole function—an arbitrary function of position. Gauge invariance is a feature of every known fundamental theory of physics. Electrodynamics, the standard model, Yang-Mills theory, gravity, all have their gauge invariances.

You may have formed the impression that gauge invariance is an interesting mathematical property with no practical significance. That would be a mistake. Gauge invariance allows us to write different but equivalent mathematical descriptions of a physical problem. Sometimes we can simplify a problem by adding something to the vector potential. For example, we can choose an S that sets any one of the components of A_μ equal to zero. Typically, we would choose a specific function S in order to illustrate or clarify one aspect of a theory. That may come at the expense of obscuring a different property of the theory. By looking at the theory from the point of view of all possible choices of S, we can see all of its properties.

Lecture 8

Maxwell's Equations

Maxwell has a private table at Hermann's Hideaway. He's sitting there alone, having a rather intense conversation with himself.

Art: *Looks like our friend Maxwell is having an identity crisis.*

Lenny: *Not exactly a crisis. He's using two different identities on purpose, to build his beautiful theory of electromagnetism.*

Art: *How far can he get with them? Identities are nice, but ...*

Lenny: *About halfway. That's where the real action begins.*

As most readers know, the Theoretical Minimum books are based on my series of lectures of the same name at Stanford University. It's in the nature of lecture series that they are not always perfectly orderly. My own lectures often begin with a review of the previous lecture or with a fill-in of some material that I didn't get to. Such was the case with Lecture 8 on Maxwell's equations. The actual lecture began not with Maxwell's equations but with the transformation properties of the electric and magnetic fields—a subject that I only hinted at, at the end of Lecture 7.

Recently I had occasion to look at Einstein's first paper on relativity.[1] I strongly encourage you to study this paper on your own. The first paragraph in particular is just marvelous. It explains his logic and motivation extremely clearly. I'd like to discuss it for a moment because it's deeply connected with the transformation properties of the electromagnetic field tensor.

8.1   Einstein's Example

Einstein's example involves a magnet and an electric conductor such as a wire. An electric current is nothing but a collection of moving charged particles. We can reduce Einstein's setup to the problem of a charged particle, contained in a wire, in the field of a magnet. The essential thing is that the wire is moving in the rest frame of the magnet.

Fig. 8.1 shows the basic setup in the "laboratory" frame where the magnet is at rest. The wire is oriented along the y axis and is moving along the x axis. An electron in the wire is dragged along with the wire. There's also a stationary

---

[1] *On the Electrodynamics of Moving Bodies*, A. Einstein, June 30, 1905.

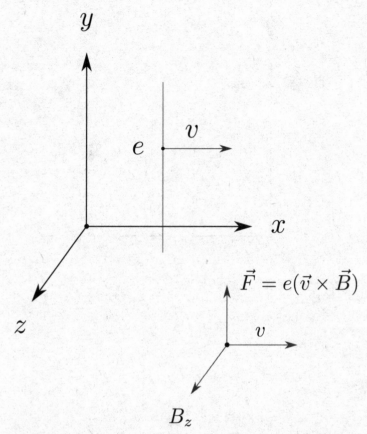

Figure 8.1: Einstein Example—Moving Charge Viewpoint. Laboratory and magnet are at rest. Charge e moves to the right at speed v. The constant magnetic field has only one component, B_z. There is no electric field.

magnet (not shown) that creates a uniform magnetic field with only one component, B_z, that points out of the page. There are no other electric or magnetic fields.

What happens to the electron? It feels a Lorentz force equal to the charge e times the cross product of its velocity v with the magnetic field B_z. That means the force will be

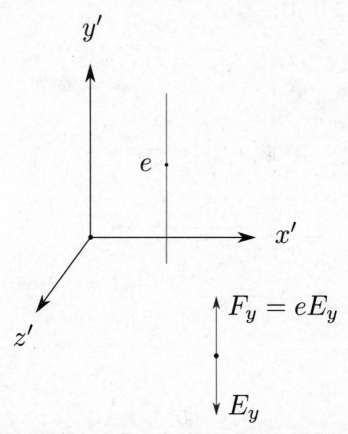

Figure 8.2: Einstein Example—Moving Magnet Viewpoint. Charge e is at rest. Magnet and laboratory move to the left at speed v. The constant electric field has only one component, E_y. The force on the electron is opposite to the direction of E_y because the electron has a negative charge.

perpendicular to both v and B_z. Following the right-hand rule, and remembering that the electron carries a negative charge, the Lorentz force will be directed upward. The result is that a current will flow in the wire. The electrons are pushed upward by the Lorentz force, but because of Benjamin Franklin's unfortunate convention (the charge of an electron is negative) the current flows downward.

That's how the laboratory observer describes the situation; the effective current is produced by the motion of the charge in a magnetic field. The small diagram at the lower right of the figure illustrates the Lorentz force

$$\vec{F} = e(\vec{v} \times \vec{B})$$

that acts on the electron.

Now let's look at the same physical situation in the frame of the electron. In this "primed" frame, the electron is at rest, and the magnet moves to the left with velocity v as shown in Fig. 8.2. Since the electron is at rest, the force on it must be due to an electric field. That's puzzling because the original problem involved only a magnetic field. The only possible conclusion is that a moving magnet must create an electric field. Boiled down to its essentials, that was Einstein's argument: The field of a moving magnet must include an electric component.

What happens when you move a magnet past a wire? A moving magnetic field creates what Einstein and his contemporaries called an electromotive force (EMF), which effectively means an electric field. Taking a magnet whose field points out of the page and moving it to the left creates a downward-oriented electric field, E_y. It exerts an upward force eE_y on a stationary electron.[2] In other words, the electric field E_y in the primed frame has the same effect as the Lorentz force in the unprimed frame. By this simple thought experiment Einstein derived the fact that magnetic fields must transform into electric fields under a Lorentz transformation. How does this prediction compare with the

---

[2] Again, the force on an electron is opposite to the electric field direction because an electron carries a negative charge.

transformation properties of $\vec{E}$ and $\vec{B}$ when they are viewed as components of an antisymmetric tensor $F_{\mu\nu}$?

8.1.1 Transforming the Field Tensor

The tensor $F_{\mu\nu}$ has components

$$F_{\mu\nu} = \begin{pmatrix} 0 & -E_x & -E_y & -E_z \\ +E_x & 0 & +B_z & -B_y \\ +E_y & -B_z & 0 & +B_x \\ +E_z & +B_y & -B_x & 0 \end{pmatrix}.$$

We want to figure out how the components transform when we move from one frame of reference to another. Given the preceding field tensor in one frame of reference, what does it look like to an observer moving in the positive direction along the x axis? What are the new components of the field tensor in the moving frame?

To work that out, we have to remember the rules for transforming tensors with two indices. Let's go back to a simpler tensor: 4-vectors, and in particular the 4-vector X^μ. 4-vectors are tensors with only one index. We know that 4-vectors transform by Lorentz transformation:

$$t' = \frac{t - vx}{\sqrt{1 - v^2}}$$

$$x' = \frac{x - vt}{\sqrt{1 - v^2}}$$

$$y' = y$$

$$z' = z.$$

As we saw in Lecture 6 (Eq. 6.3, for example), we can

organize these equations into a matrix expression using the Einstein summation convention. We introduced the matrix $L^\mu{}_\nu$ that captures the details of a simple Lorentz transformation along the x axis. Using this notation, we wrote the transformation as

$$(X')^\mu = L^\mu{}_\nu X^\nu,$$

where ν is a summation index. This is just shorthand for the preceding set of four Lorentz transformation equations. In Eq. 6.4, we saw what this equation looks like in matrix form. Here's a slightly revised version of that equation, with (t, x, y, z) replaced by (X^0, X^1, X^2, X^3):

$$
\begin{pmatrix} X^0 \\ X^1 \\ X^2 \\ X^3 \end{pmatrix}' = \begin{pmatrix} \dfrac{1}{\sqrt{1-v^2}} & \dfrac{-v}{\sqrt{1-v^2}} & 0 & 0 \\ \dfrac{-v}{\sqrt{1-v^2}} & \dfrac{1}{\sqrt{1-v^2}} & 0 & 0 \\ 0 & 0 & 1 & 0 \\ 0 & 0 & 0 & 1 \end{pmatrix} \begin{pmatrix} X^0 \\ X^1 \\ X^2 \\ X^3 \end{pmatrix}. \tag{8.1}
$$

How do we transform a tensor with two indices? For this example, I'll use the upper-index version of the field tensor,

$$
F^{\mu\nu} = \begin{pmatrix} 0 & +E_x & +E_y & +E_z \\ -E_x & 0 & +B_z & -B_y \\ -E_y & -B_z & 0 & +B_x \\ -E_z & +B_y & -B_x & 0 \end{pmatrix}.
$$

I've chosen the upper-index version (Eq. 6.42) for convenience, because we already have rules to transform things with upper indices; the form of the Lorentz transformation we've been using all along is set up to operate on things with upper indices. How do you transform a thing with *two* upper indices? It's almost the same as transforming a single-index thing. Where the single-index transformation rule is

$$(X')^\mu = L^\mu{}_\sigma X^\sigma,$$

with summation index σ, the two-index transformation rule is

$$(F')^{\mu\nu} = L^\mu{}_\sigma L^\nu{}_\tau F^{\sigma\tau}, \tag{8.2}$$

with two summation indices σ and τ. In other words, we just perform the single-index transformation twice. Each index of the original object $F^{\sigma\tau}$ becomes a summation index in the transformation equation. Each index (σ and τ) transforms in exactly the same way that it would if it were the *only* index being transformed.

If you had an object with more indices—you could have any number of indices—the rule would be the same: You transform every index in the same way. That's the rule for transforming tensors. Let's try it out with $F^{\mu\nu}$ and see if we can discover whether there is an electric field along the y axis. In other words, let's try to compute the y component of electric field $(E')^y$ in the primed reference frame.

$$(E')^y = (F')^{0y}$$

What do we know about the original unprimed field tensor $F^{\mu\nu}$ as it applies to Einstein's example? We know that it represents a pure magnetic field along the *unprimed z* axis. Therefore, $F^{\sigma\tau}$ has only one nonzero component, F^{xy}, which corresponds to B^z.[3] Following the pattern of Eq. 8.2, we can write

$$(E')^y = (F')^{0y} = L^0{}_x L^y{}_y F^{xy}. \tag{8.3}$$

---

[3] The antisymmetric component $F^{yx} = -F^{xy}$ is also nonzero, but this adds no new information.

Notice that the indices match up in the same way as the indices in Eq. 8.2, even though we're only looking at one single component on the left side. The summations on the right collapse to a single term. By referring to the matrix in Eq. 8.1, we can identify the specific elements of L to substitute into the transformation equation 8.3. Specifically, we can see that

$$L^0_{\ x} = \frac{-v}{\sqrt{1 - v^2}}$$

and

$$L^y_{\ y} = 1.$$

Applying these substitutions in Eq. 8.3 results in

$$(E')^y = (F')^{0y} = \frac{-v}{\sqrt{1 - v^2}}(1)F^{xy},$$

or

$$(E')^y = (F')^{0y} = \frac{-vF^{xy}}{\sqrt{1 - v^2}}.$$

But F^{xy} is the same as B^z in the original reference frame, and we can now write

$$(E')^y = (F')^{0y} = \frac{-vB^z}{\sqrt{1 - v^2}}. \tag{8.4}$$

Thus, as Einstein claimed, a pure magnetic field when viewed from a moving frame will have an electric field component.

Exercise 8.1: Consider an electric charge at rest, with no additional electric or magnetic fields present. In terms of the rest frame components, (E_x, E_y, E_z), what is the x component of the electric field for an observer moving in the negative x direction with velocity v? What are the y and z components? What are the corresponding components of the magnetic field?

Exercise 8.2: Art is sitting in the station as the train passes by. In terms of Lenny's field components, what is the x component of E observed by Art? What are the y and z components? What are the components of the magnetic field seen by Art?

8.1.2   Summary of Einstein's Example

We started out by setting up a laboratory frame with an electric field of zero, a magnetic field pointing only in the positive z direction, and an electron moving with velocity v in the positive x direction. We then asked for the electric and magnetic field values in the frame of the electron. We discovered two things:

1. There's no magnetic force on the electron in the new frame because the electron is at rest.
2. There is, however, a y component of electric field in the new reference frame that exerts a force on the electron.

The force on the (moving) electron that is due to a magnetic field in the laboratory frame is due to an electric field in the moving frame (the electron's rest frame). That's the essence of Einstein's example. Now, on to Maxwell's equations.

> **Exercise 8.3:** For Einstein's example, work out all components of the electric and magnetic fields in the electron's rest frame.

8.2 Introduction to Maxwell's Equations

Recall Pauli's (fictional) quote:

> *Fields tell charges how to move; charges tell fields how to vary.*

In Lecture 6 we spent a good deal of time on the first half of the quote, describing the way fields tell charges how to move. Now it's the charges' turn to tell fields how to vary. If fields control particles through the Lorentz force law, charges control fields through Maxwell's equations.

My philosophy of teaching electrodynamics is somewhat unorthodox, so let me explain it. Most courses take a historical perspective, beginning with a set of laws that were discovered in the late eighteenth and early nineteenth centuries. You've probably heard of them—Coulomb's law, Ampère's law, and Faraday's law. These laws are sometimes taught without calculus, a grave mistake in my opinion; physics is always harder without the mathematics. Then, through a somewhat torturous procedure, these laws are manipulated into Maxwell's equations. My own way of teaching, even to undergraduates, is to take the plunge— from the start—and begin with Maxwell's equations. We take each of the equations, analyze its meaning, and in that way derive Coulomb's, Ampère's and Faraday's laws. It's the "cold shower" way of doing it, but within a week or two the students understand what would take months doing it the historical way.

How many Maxwell's equations are there altogether? Actually there are eight, although vector notation boils them down to four: two 3-vector equations (with three components apiece) and two scalar equations.

Four of the equations are identities that follow from definitions of the electric and magnetic fields in terms of the vector potential. Specifically, they follow directly from the definitions

$$F_{\mu\nu} = \partial_\mu A_\nu - \partial_\nu A_\mu$$
$$E_n = -(\partial_0 A_n - \partial_n A_0)$$
$$\vec{B} = \vec{\nabla} \times \vec{A},$$

together with some vector identities. There's no need to invoke an action principle. In that sense, this subset of Maxwell's equations comes "for free."

8.2.1 Vector Identities

An identity is a mathematical fact that follows from a definition. Many are trivial, but the interesting ones are often not obvious. Over the years, people have stockpiled a large number of identities involving vector operators. For now, we need only two. You can find a summary of basic vector operators in Appendix B.

Two Identities

Now for our two identities. The first one says that the divergence of a curl is always zero. In symbols,

$$\vec{\nabla} \cdot (\vec{\nabla} \times \vec{A}) = 0$$

for any vector field $\vec{A}$. You can prove this by a technique known as "brute force." Just expand the entire left side in component form, using the definitions of divergence and curl (Appendix B). You'll find that many of the terms cancel. For example, one term might involve the derivative of something with respect to x and then with respect to y, while another term involves the same derivatives in the opposite order, and with a minus sign. Terms of that kind cancel out to zero.

The second identity states that the curl of any gradient is zero. In symbols,

$$\vec{\nabla} \times (\vec{\nabla} S) = 0,$$

where S is a scalar and $\vec{\nabla} S$ is a vector. Remember, these two identities are not statements about some special field like the vector potential. They're true for *any* fields S and $\vec{A}$, as long as they're differentiable. These two identities are enough to prove half of Maxwell's equations.

8.2.2 Magnetic Field

We saw in Eq. 6.30 that the magnetic field $\vec{B}$ is the curl of the vector potential $\vec{A}$,

$$\vec{B} = \vec{\nabla} \times \vec{A}.$$

Therefore, the divergence of a magnetic field is the divergence of a curl,

$$\vec{\nabla} \cdot \vec{B} = \vec{\nabla} \cdot (\vec{\nabla} \times \vec{A}).$$

Our first vector identity tells us that the right side must be zero. In other words,

$$\vec{\nabla} \cdot \vec{B} = 0. \tag{8.5}$$

This is one of Maxwell's equations. It says that there cannot be magnetic charges. If we saw a configuration with magnetic field vectors diverging from a single point—a magnetic monopole—then this equation would be wrong.

8.2.3   Electric Field

We can derive another Maxwell equation—a vector equation involving the electric field—from vector identities. Let's go back to the definition of the electric field,

$$E_n = -\left(\frac{\partial A_n}{\partial t} - \frac{\partial A_0}{\partial X^n} \right). \tag{8.6}$$

Notice that the second term is just the nth component of the gradient of A_0. From a three-dimensional point of view (just space), the time component of $\vec{A}$, that is, A_0, can be thought of as a scalar.[4] The vector whose components are derivatives of A_0 can be thought of as the gradient of A_0. So the electric field has two terms; one is the time derivative of the space component of $\vec{A}$, and the other is the gradient of the time component. We can rewrite Eq. 8.6 as a vector equation,

$$\vec{E} = -\left(\frac{\partial \vec{A}}{\partial t} - \vec{\nabla} A_0 \right). \tag{8.7}$$

Now let's consider the curl of $\vec{E}$,

$$\vec{\nabla} \times \vec{E} = -\vec{\nabla} \times \frac{\partial \vec{A}}{\partial t} + \vec{\nabla} \times \vec{\nabla} A_0. \tag{8.8}$$

The second term on the right side is the curl of a gradient, and our second vector identity tells us that it must be zero.

---

[4] It's not a scalar from the point of view of four-dimensional vector spaces.

We can rewrite the first term as

$$-\vec{\nabla} \times \frac{\partial \vec{A}}{\partial t} = -\frac{\partial}{\partial t}(\vec{\nabla} \times \vec{A}).$$

Why is that? Because we're allowed to interchange derivatives. When we take a derivative to make a curl, we can interchange the space derivative with the time derivative, and pull the time derivative on the outside. So the curl of the time derivative of $\vec{A}$ is the time derivative of the curl of $\vec{A}$. As a result, Eq. 8.8 simplifies to

$$\vec{\nabla} \times \vec{E} = -\frac{\partial}{\partial t}(\vec{\nabla} \times \vec{A}).$$

But we already know (Eq. 6.30) that the curl of $\vec{A}$ is the magnetic field $\vec{B}$. So we can write a second Maxwell equation,

$$\vec{\nabla} \times \vec{E} = -\frac{\partial}{\partial t}\vec{B} \tag{8.9}$$

or

$$\vec{\nabla} \times \vec{E} + \frac{\partial}{\partial t}\vec{B} = 0. \tag{8.10}$$

This vector equation represents three equations, one for each component of space. Eqs. 8.5 and 8.10 are the so-called *homogeneous* Maxwell equations.

Although the homogeneous Maxwell equations were derived as identities, they nevertheless have important content. Equation 8.5,

$$\vec{\nabla} \cdot \vec{B} = 0, \tag{8.11}$$

expresses the important fact that there are no magnetic charges in nature, if indeed it is a fact. In elementary terms it states that, unlike electric fields, the lines of magnetic flux never end. Does that mean that magnetic monopoles—

magnetic analogs of electrically charged particles—are impossible? I won't answer that here but at the end of the book we'll examine the issue in some detail.

What about Eq. 8.10? Does it have consequences? Certainly it does. Equation 8.10 is one mathematical formulation of Faraday's law that governs the workings of all electro-mechanical devices such as motors and generators. At the end of this lecture we'll come back to Faraday's law among others.

8.2.4 Two More Maxwell Equations

There are two more Maxwell equations[5] that are not mathematical identities, which means that they cannot be derived just from the definitions of $\vec{E}$ and $\vec{B}$. It will be our eventual goal to derive them from an action principle, but originally they were discovered as empirical laws that summarized the results of experiments on charges, electric currents, and magnets. The two additional equations are similar to Eqs. 8.5 and 8.10, with the electric and magnetic fields switching roles. The first is a single equation,

$$\vec{\nabla} \cdot \vec{E} = \rho. \qquad (8.12)$$

This is similar to Eq. 8.5 except that the right side is not zero. The quantity ρ is the density of electric charge, that is, the charge per unit volume at each point of space. The equation represents the fact that electric charges are surrounded by electric fields that they carry with them wherever they go. It is closely connected to Coulomb's law, as we will see.

The second equation—really three equations—is similar

---

[5] There are four more if you write them in component form.

Derivation	Equation
From Vector Identities:	$\vec{\nabla} \cdot \vec{B} = 0$
(Homogeneous Equations)	$\vec{\nabla} \times \vec{E} + \dfrac{\partial}{\partial t}\vec{B} = 0$
From Action Principle:	$\vec{\nabla} \cdot \vec{E} = \rho$
(Inhomogeneous Equations)	$\vec{\nabla} \times \vec{B} - \dfrac{\partial}{\partial t}\vec{E} = \vec{j}$

Table 8.1: Maxwell Equations and Their Derivations from the Vector Potential.

to Eq. 8.10:

$$\vec{\nabla} \times \vec{B} - \frac{\partial}{\partial t}\vec{E} = \vec{j}. \tag{8.13}$$

Again, aside from the interchange of electric and magnetic fields and a change of sign, the most important difference between Eq. 8.10 and Eq. 8.13 is that the right side of Eq. 8.13 is not zero. The quantity $\vec{j}$ is the current density, which represents the flow of charge—for example, the flow of electrons through a wire.

The content of Eq. 8.13 is that flowing currents of charge are also surrounded by fields. This equation is also related to a law, Ampère's law, that determines the magnetic field produced by a current in a wire.

Table 8.1 shows all four Maxwell equations. We derived the first two by applying vector identities to the vector potential. The second two equations also come from the vector potential, but they contain dynamical information and we need an action principle to derive them. We'll do that in the next lecture.

The second group of equations looks a lot like the first group, but with the roles of $\vec{E}$ and $\vec{B}$ (almost) interchanged. I say *almost* because there are some small but important differences in their structure. To begin with, the signs are a little different. Also, the equations in the second group are *inhomogeneous*; they involve quantities on the right side called *charge density* ρ and *current density* $\vec{j}$ that do not appear in the first group. We'll have more to say about ρ and $\vec{j}$ in the next section. Charge density ρ is the amount of charge per unit volume, just as you would expect; in three dimensions, it's a scalar. Current density is a 3-vector. We'll see in a moment that in 4-vector language things are a little more complicated. We'll discover that ρ together with the three components of $\vec{j}$ are the components of a 4-vector.[6]

8.2.5   Charge Density and Current Density

What are charge density and current density? I've tried to illustrate them in Figs. 8.3 and 8.4, but there's a problem: I don't know how to draw four dimensions of spacetime on a two-dimensional page. Keep that in mind as you interpret these figures. They require some abstract thinking.

Charge density is exactly what it sounds like. You look at a small region of three-dimensional space and divide the total amount of charge in that region by the region's volume. Charge density is the limiting value of this quotient as the volume becomes very small.

Let's describe this idea from the perspective of four-dimensional spacetime. Fig. 8.3 tries to depict all four directions of spacetime. As usual, the vertical axis represents time, and the x axis points to the right. The spatial axis pointing out of the page is labeled y, z and represents both the y and z

---

[6] In four-dimensional spacetime, therefore, ρ is not a scalar. It transforms as the time component of a 4-vector.

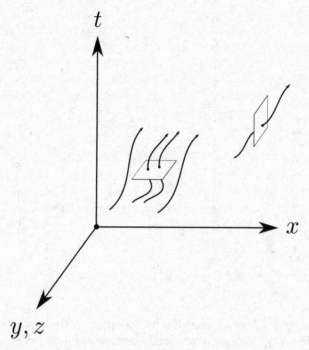

Figure 8.3: Charge and Current Density in Spacetime. The y, z axis represents two directions in space. Curved arrows are world lines of charged particles. The horizontal window illustrates charge density. The vertical window illustrates the x component of current density.

directions. This is the "abstract thinking" part of the diagram; it's far from perfect, but you get the idea. Let's consider a little cell in space, shown in the figure as a horizontal square. There are two important things to notice about it. First, its orientation is perpendicular to the time axis. Second, it's not really a square at all. The "edges" parallel to the y, z axis are really areas, and the square is

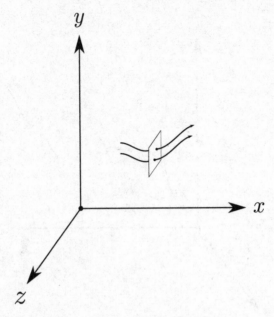

Figure 8.4: Current Density in Space. The time axis is not shown. The curved arrows are not world lines, but show trajectories in space only.

really a three-dimensional volume element, in other words, a cube.

The curved arrows represent world lines of charged particles. We can imagine that space and time are filled with these world lines, flowing like a fluid. Consider the world lines that pass through our little cube. Remember that our cube, being perpendicular to the time axis, represents a single instant of time. This means that the particles that "pass through" the cube are simply the particles that are *in* the cube at that instant of time. We count up all their charges to get the total charge passing through that little volume of space.[7] The charge density ρ is the limiting value of the total charge in the volume element divided by its volume,

$$\rho = \frac{\Delta Q}{\Delta V}.$$

What about current density? One way to think about current density is to turn our little square onto its edge. The vertically oriented "square" is shown to the right in Fig. 8.3. This diagram calls for even more abstract thinking. The square still represents a three-dimensional cube in spacetime. But this cube is not purely spatial; because one edge is parallel to the t axis, one of its three dimensions is time. The edge parallel to the y, z axis confirms that the other two dimensions are spatial. This square is perpendicular to the x axis. As before, the edges parallel to the y, z axis really stand for two-dimensional squares in the y, z plane. Let's re-draw this little square in Fig. 8.4 as a pure space diagram with only the x, y, and z axes. In this new diagram the square really is a square because there's no time axis. It's just a little window perpendicular to the x axis. How much charge passes through that window per unit time? The window is an area, so we're really asking how much charge flows through the window per unit area per unit time.[8] That's what we mean by current density. Because our window is perpendicular to the x axis, it defines current density in the x direction, and we have similar components for the y and z directions. Current density $\vec{j}$ is a

---

[7] *Passing through* has a particular meaning in this context. The volume element exists at one instant of time. The particles were in the past, and then they're in the future. But at the instant represented by our volume element, they are *in* the volume element.

[8] That's the precise analog of turning the charge density rectangle on its edge in Fig. 8.3.

3-vector,

$$j_x = \frac{\Delta Q}{\Delta A_x \Delta t}$$

$$j_y = \frac{\Delta Q}{\Delta A_y \Delta t}$$

$$j_z = \frac{\Delta Q}{\Delta A_z \Delta t}. \tag{8.14}$$

The quantities ΔA_x, ΔA_y, and ΔA_z represent elements of area that are perpendicular to the x, y, and z axes respectively. In other words (see Fig. 8.6), the ΔA's that appear in Eqs. 8.14 are

$$\Delta A_x = \Delta y \Delta z$$
$$\Delta A_y = \Delta z \Delta x$$
$$\Delta A_z = \Delta x \Delta y,$$

and therefore we can write Eqs. 8.14 in the form

$$j_x = \frac{\Delta Q}{\Delta y \Delta z \Delta t}$$

$$j_y = \frac{\Delta Q}{\Delta z \Delta x \Delta t}$$

$$j_z = \frac{\Delta Q}{\Delta x \Delta y \Delta t}. \tag{8.15}$$

You can think of ρ itself as a flow of charge in the time direction, the x component of $\vec{j}$ as a flow in the x direction, and so forth. You can think of a spatial volume as a kind of window perpendicular to the t axis. We'll soon discover that the quantities (ρ, j_x, j_y, j_z) are actually the contravariant components of a 4-vector. They enter into the other two Maxwell equations, the equations that we'll derive from an action principle in the next lecture.

8.2.6 Conservation of Charge

What does charge conservation actually mean? One of the things it means is that the total amount of charge never changes; if the total amount of charge is Q, then,

$$\frac{dQ}{dt} = 0. \qquad (8.16)$$

But that's not all. Suppose Q could suddenly disappear from our laboratory on Earth and instantly reappear on the moon (Fig. 8.5). A physicist on Earth would conclude that Q is not conserved. Conservation of charge means something stronger than "the total amount of Q does not change." It really means that whenever Q decreases (or increases) within the walls of the laboratory, it increases (or decreases) by the same amount right outside the walls. Even better, a change in Q is accompanied by a flux of Q passing *through* the walls. When we say *conservation*, what we really mean is *local conservation*. This important idea applies to other conserved quantities as well, not just charge.

To capture the idea of local conservation mathematically, we need to define symbols for flux, or flow across a boundary. The terms *flux*, *flow*, and *current* of a quantity are synonymous and refer to the amount flowing through a small element of area per unit area per unit time. Changes in the amount of Q in a region of space must be accounted for by a flux of Q through the boundaries of the region.

Fig. 8.6 shows a three-dimensional volume element that contains a charge Q. The walls of the box are little windows.

Figure 8.5: Local Conservation of Charge. The process in this picture cannot happen. Charge can't disappear in one place and instantly reappear in some far-removed place.

Any change in the amount of charge inside the box must be accompanied by current through its boundaries. For convenience, let's assume the volume of the box has a value of 1 in some units where a small volume is just a unit volume. Similarly, we can take the edges of the box, $(\Delta x, \Delta y, \Delta z)$, to have unit length.

What is the time derivative of the charge inside the box? Because the box has unit volume, the charge inside the box is the same as the charge density ρ. Therefore, the time derivative of the charge is just $\dfrac{\partial \rho}{\partial t}$, the time derivative of ρ. Local charge conservation says that $\dfrac{\partial \rho}{\partial t}$ must equal the amount of charge flowing into the box through its boundaries. For example, suppose the charge Q increases. That means there must be some net inflow of charge.

The net inflow of charge must be the sum of the charges coming in through each of the six windows (the six faces of the cube). Let's consider the charge coming in through the right-facing window, labeled $-j_{x+}$ in the diagram. The $x+$

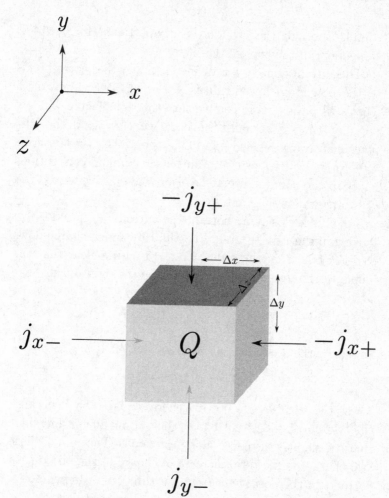

Figure 8.6: Local Charge Conservation. For convenience, we assume $\Delta x = \Delta y = \Delta z = 1$ in some small units. The volume element is $\Delta V = \Delta x \Delta y \Delta z = 1$.

subscript means the x component of current that enters the right-facing or + window of the box. This notation seems fussy, but in a moment you'll see why we need it. The charge coming through that window is proportional to j_x, the x component of the current density.

The current density j_x is defined in a way that corresponds to the flow from lower values of x to greater values of x. That's why the charge coming into the right-facing window per unit time is actually the negative of j_x, and the left-pointing arrow is labeled $-j_{x+}$. Likewise, we use the subscript "$x-$" for the leftmost or minus face. Similar conventions apply to the y and z directions. To avoid clutter, we did not draw arrows for $-j_{z+}$ and j_{z-}.

Let's look closely at both of the currents (j_{x-} and $-j_{x+}$) flowing in the x direction in Fig. 8.6. How much charge flows into the box from the right? Eq. 8.15 makes clear that the amount of charge entering the box from the right during time Δt is

$$\Delta Q_{right} = -(j_{x+})\Delta y \Delta z \Delta t.$$

Similarly, the amount entering from the left is

$$\Delta Q_{left} = (j_{x-})\Delta y \Delta z \Delta t.$$

These two charges are not in general equal. That's because the two currents are not equal; they occur at two locations that are slightly displaced along the x axis. The sum of these two charges is the total increase of charge in the box due to currents in the positive or negative x direction. In symbols,

$$\Delta Q_{total} = -(j_{x+})\Delta y \Delta z \Delta t + (j_{x-})\Delta y \Delta z \Delta t$$

or

$$\Delta Q_{total} = -(j_{x+} - j_{x-})\Delta y \Delta z \Delta t. \qquad (8.17)$$

The term in parentheses is the change in current over a small interval Δx and is closely related to the derivative of j_x with respect to x. In fact, it's

$$-(j_{x+} - j_{x-}) = -\frac{\partial j_x}{\partial x}\Delta x.$$

Substituting that back into Eq. 8.17 results in

$$\Delta Q_{total} = -\frac{\partial j_x}{\partial x}\Delta x \Delta y \Delta z \Delta t. \qquad (8.18)$$

We easily recognize $\Delta x \Delta y \Delta z$ as the volume of the box. Dividing both sides by the volume gives

$$\frac{\Delta Q_{total}}{\Delta x \Delta y \Delta z} = -\frac{\partial j_x}{\partial x}\Delta t. \qquad (8.19)$$

The left side of Eq. 8.19 is a change in charge per unit volume. In other words, it's a change in the charge density ρ, and we can write

$$\Delta\rho_{total} = -\frac{\partial j_x}{\partial x}\Delta t. \qquad (8.20)$$

Dividing by Δt, and taking the limit as small quantities approach zero,

$$\frac{\partial\rho_{total}}{\partial t} = -\frac{\partial j_x}{\partial x}. \qquad (8.21)$$

Now Eq. 8.21 is not really correct; it only accounted for the charge coming into or out of the box through the two walls or perpendicular to the x axis. To put it another way, the subscript "total" in this equation is misleading because it only accounts for current densities in that one direction. To get the whole picture, we would run the same analysis for the y and z directions and then add up the contributions from current densities in all three directions. The result is

$$\frac{\partial \rho}{\partial t} = -\left(\frac{\partial j_x}{\partial x} + \frac{\partial j_y}{\partial y} + \frac{\partial j_z}{\partial z}\right). \tag{8.22}$$

The right side is the divergence of $\vec{j}$, and in vector notation Eq. 8.22 becomes

$$\frac{\partial \rho}{\partial t} = -\vec{\nabla} \cdot \vec{j}$$

or

$$\frac{\partial \rho}{\partial t} + \vec{\nabla} \cdot \vec{j} = 0. \tag{8.23}$$

This equation is a local neighborhood-by-neighborhood expression of the conservation of charge. It says that the charge within any small region will only change due to the flow of charge through the walls of the region. Eq. 8.23 is so important that it has a name: it's called the *continuity equation*. I don't like this name because the word *continuity* gives the impression that the charge distribution needs to be continuous, and it does not. I prefer to call it *local conservation*. It describes a strong form of conservation that does not allow a charge to disappear at one place and reappear at another place unless there's a flow of charge (a current) in between.

We could add this new equation to Maxwell's equations, but in fact we don't need to. Eq. 8.23 is actually a consequence of Maxwell's equations. The proof is not difficult; it's a fun exercise that I'll leave to you.

Exercise 8.4: Use the second group of Maxwell's equations from Table 8.1 along with the two vector identities from Section 8.2.1 to derive the continuity equation.

8.2.7 Maxwell's Equations: Tensor Form

By now, we've seen all four of Maxwell's equations in 3-vector form (Table 8.1). We have not yet derived the second group of these equations from an action principle, but we'll do that in Lecture 10. Before we do, I'd like to show you how to write the first group of equations (along with the continuity equation) in tensor notation with upstairs and downstairs indices. This will make clear that the equations are invariant under Lorentz transformations. In this section, we'll switch back to 4-vector notation that follows the rules of indexology. For example, I'll start writing (J^x, J^y, J^z) or (J^1, J^2, J^3) for the space components of current density.

Let's start with Eq. 8.23, the continuity equation. We can rewrite it as

$$\frac{\partial \rho}{\partial t} + \frac{\partial J^x}{\partial x} + \frac{\partial J^y}{\partial y} + \frac{\partial J^z}{\partial z} = 0.$$

Now let's define ρ to be the time component of J:

$$\rho = J^0.$$

In other words, take (J^x, J^y, J^z) and put them together with ρ to form a complex of four numbers. We'll call this new object J^μ:

$$J^\mu = (\rho, J^x, J^y, J^z).$$

Using this notation, we can rewrite the continuity equation as

$$\frac{\partial J^\mu}{\partial X^\mu} = 0. \tag{8.24}$$

Notice that taking the derivative with respect to X^μ does two things: First, it adds a covariant index. Second, because the new covariant index is the same as the contravariant index in

J^μ, it triggers the summation convention. Written in this way, the continuity equation has the appearance of a 4-dimensional scalar equation. It's not hard to prove that the components of J^μ really do transform as the components of a 4-vector. Let's do that now.

To begin with, it's a well-verified experimental fact that electric charge is Lorentz invariant. If we measure a charge to be Q in one frame, it will be Q in every frame. Now let's revisit the small charge ρ inside our little unit-volume box. Suppose this box and its enclosed charge are moving to the right with velocity v in the laboratory frame.[9] What are the four components of current density in the *rest frame of the box*? Clearly, they must be

$$(J')^\mu = (\rho, 0, 0, 0),$$

because the charge is standing still in this frame. What are the components J^μ in the laboratory frame? We claim they're

$$J^\mu = \left(\frac{\rho}{\sqrt{1 - v^2}}, \frac{\rho v}{\sqrt{1 - v^2}}, 0, 0 \right),$$

or equivalently,

$$J^\mu = \rho \left(\frac{1}{\sqrt{1 - v^2}}, \frac{v}{\sqrt{1 - v^2}}, 0, 0 \right). \tag{8.25}$$

To see this, remember that charge density is determined by two factors: the amount of charge and the size of the region that contains the charge. We already know that the amount of charge is the same in both frames. What about the volume of the enclosing region, namely the box? In the laboratory frame, the moving box undergoes the standard Lorentz

---

[9] As usual, the moving frame has primed coordinates, and laboratory frame is unprimed.

contraction, $\sqrt{1 - v^2}$. Because the box contracts only in the x direction, its volume is reduced by the same factor. But density and volume are inversely related, so a decrease in volume causes a reciprocal increase in density. That explains why the components on the right side of Eq. 8.25 are correct. Finally, notice that these components are actually the components of 4-velocity, which is a 4-vector. In other words, J^μ is the product of an invariant scalar quantity ρ with a 4-vector. Therefore J^μ itself must also be a 4-vector.

8.2.8 The Bianchi Identity

Let's take stock of what we know: $\vec{E}$ and $\vec{B}$ form an antisymmetric tensor $F_{\mu\nu}$ with two indices. The current density 3-vector $\vec{j}$ together with ρ form a 4-vector, that is, a tensor with one index. We also have the first group of Maxwell equations from Table 8.1,

$$\vec{\nabla} \cdot \vec{B} = 0 \tag{8.26}$$

$$\vec{\nabla} \times \vec{E} + \frac{\partial}{\partial t}\vec{B} = 0. \tag{8.27}$$

These equations are a consequence of the definitions of $\vec{B}$ and $\vec{E}$ in terms of the vector potential. Each of them has the form of partial derivatives acting on components of $F_{\mu\nu}$, with a result of zero. In other words, they have the form

$$\partial F = 0.$$

I'm using the symbol ∂ loosely, to mean "some combination of partial derivatives." Eqs. 8.26 and 8.27 represent four equations altogether: one corresponding to Eq. 8.26 because it equates two scalars, and three corresponding to Eq. 8.27 because it equates two 3-vectors. How can we write these equations in Lorentz invariant form? The easiest approach is to just write down the answer and then explain why it works. Here it is:

$$\partial_\sigma F_{\nu\tau} + \partial_\nu F_{\tau\sigma} + \partial_\tau F_{\sigma\nu} = 0. \qquad (8.28)$$

Eq. 8.28 is called the Bianchi identity.[10] The indices σ, ν, and τ can take on any of the four values $(0, 1, 2, 3)$ or (t, x, y, z). No matter which of these values we assign to σ, ν, and τ, Eq. 8.28 gives a result of zero.

Let's convince ourselves that the Bianchi identity is equivalent to the two homogeneous Maxwell equations. First, consider the case where all three indices σ, ν, and τ represent space components. In particular, suppose we choose them to be

$$\sigma = y$$

$$\nu = x$$

$$\tau = z.$$

With a little algebra (recall that $F_{\mu\nu}$ is antisymmetric), substituting these index values into Eq. 8.28 gives

$$\partial_x F_{yz} + \partial_y F_{zx} + \partial_z F_{xy} = 0.$$

But the pure space components of F, such as F_{yx}, correspond to the components of the magnetic field. Recall (Eq. 6.41) that

$$F_{yz} = B_x$$

$$F_{zx} = B_y$$

$$F_{xy} = B_z.$$

---

[10] It's actually a special case of the Bianchi identity.

Substituting those values, we can write

$$\partial_x B_x + \partial_y B_y + \partial_z B_z = 0.$$

The left side is just the divergence of $\vec{B}$, so this equation is equivalent to Eq. 8.26. It turns out that if we choose a different way of assigning x, y, and z to the Greek indices, we get the same result. Go ahead and try it.

What happens if one of the indices is a time component? Let's try the assignment

$$\sigma = y$$

$$\nu = x$$

$$\tau = t,$$

resulting in

$$\partial_y F_{xt} + \partial_x F_{ty} + \partial_t F_{yx} = 0.$$

Two of the three terms have mixed space and time components—one leg in time and one leg in space. That means they're components of the electric field. Recall (Eq. 6.41) that

$$F_{xt} = E_x$$
$$F_{ty} = -E_y$$
$$F_{yx} = -B_z.$$

The substitution results in

$$\partial_y E_x - \partial_x E_y - \partial_t B_z = 0.$$

If we flip the sign (multiply by -1), this becomes

$$\partial_x E_y - \partial_y E_x + \partial_t B_z = 0.$$

This is just the z component of Eq. 8.27, which says that the curl of $\vec{E}$ plus the time derivative of $\vec{B}$ is zero. If we tried other combinations with one time component and two space components, we would discover the x and y components of Eq. 8.27.

How many ways are there to assign values to the indices σ, ν, and τ? There are three indices, and each index can be assigned one of the four values t, x, y, and z. That means there are $4 \times 4 \times 4 = 64$ different ways to do it. How can it be that the sixty-four equations of the Bianchi identity are equivalent to the four homogeneous Maxwell equations? It's simple; the Bianchi identity includes many equations that are redundant. For example, any index assignment where two indices are equal will result in the trivial equation $0 = 0$. Also, many of the index assignments are redundant; they will yield the same equation. Every one of the nontrivial equations is equivalent either to Eq. 8.26 or to one of the components of Eq. 8.27.

There's another way to check the Bianchi identity. Remember that $F_{\mu\nu}$ is defined to be

$$F_{\mu\nu} = \partial_\mu A_\nu - \partial_\nu A_\mu,$$

or equivalently,

$$F_{\mu\nu} = \frac{\partial A_\nu}{\partial X^\mu} - \frac{\partial A_\mu}{\partial X^\nu}.$$

By making the appropriate substitutions into Eq. 8.28, we can write it as

$$\partial_\sigma \left(\frac{\partial A_\tau}{\partial X^\nu} - \frac{\partial A_\nu}{\partial X^\tau} \right) + \partial_\nu \left(\frac{\partial A_\sigma}{\partial X^\tau} - \frac{\partial A_\tau}{\partial X^\sigma} \right) + \partial_\tau \left(\frac{\partial A_\nu}{\partial X^\sigma} - \frac{\partial A_\sigma}{\partial X^\nu} \right) = 0.$$

If you expand these derivatives, you'll discover that they cancel term by term, and the result is zero. Eq. 8.28 is completely Lorentz invariant. You can check that by looking at how it transforms.

Lecture 9

Physical Consequences of Maxwell's Equations[1]

Art: *Lenny, is that Faraday sitting over there by the window?*

Lenny: *Yes, I think so. You can tell because his table is not cluttered with fancy equations. Faraday's results—and there are plenty of them—come straight from the laboratory.*

Art: *Well, I like equations, even when they're hard. But I also wonder if we need them as much as we think. Just look what Faraday accomplishes without their help!*

Just then, Faraday notices Maxwell across the room. They exchange a friendly wave.

---

[1] Somehow, this important lecture never made it to the video site for the course. It represents the main point of contact between Lenny's approach to the subject and the traditional approach. As a result of this insertion, Lecture 10 in the book corresponds to Lecture 9 in the video series, and Lecture 11 in the book corresponds to Lecture 10 in the video series.

9.1   Mathematical Interlude

The fundamental theorem of calculus relates derivatives and integrals. Let me remind you of what it says. Suppose we have a function $F(x)$ and its derivative dF/dx. The fundamental theorem can be stated simply:

$$\int_a^b \frac{dF}{dx}\, dx = F(b) - F(a). \tag{9.1}$$

Notice the form of Eq. 9.1: On the left side we have an integral in which the integrand is a derivative of $F(x)$. The integral is taken over an interval from a to b. The right side involves the values of F at the boundaries of the interval. Eq. 9.1 is the simplest case of a far more general type of relation, two famous examples being *Gauss's theorem* and *Stokes's theorem*. These theorems are important in electromagnetic theory, and so I will state and explain them here. I will not prove them, but you can easily find proofs in many places, including the Internet.

9.1.1   Gauss's Theorem

Instead of a one-dimensional interval $a < x < b$, let's consider a region of three-dimensional space. The interior of a sphere or a cube would be examples. But the region does not have to be regularly shaped. Any blob of space will do. Let's call the region B for *blob*.

The blob has a boundary or surface that we will call S for *surface*. At each point on the surface we can construct a unit vector $\hat{n}$ pointing outward from the blob. All of this is shown in Figure 9.1. By comparison to the one-dimensional analogy (Eq. 9.1), the blob would replace the interval between a and b, and the boundary surface S would replace the two points a and b.

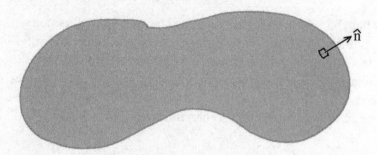

Figure 9.1: Illustration of Gauss's Theorem. $\hat{n}$ is an outward-pointing unit normal vector.

Instead of a simple function F and its derivative dF/dx, we will consider a vector field $\vec{V}(x, y, z)$ and its divergence $\vec{\nabla} \cdot \vec{V}$. In analogy with Eq. 9.1, Gauss's theorem asserts a relation between the integral of $\vec{\nabla} \cdot \vec{V}$ over the three-dimensional blob of space B, and the values of the vector field on the two-dimensional boundary of the blob, S. I will write it and then explain it:

$$\int_B \vec{\nabla} \cdot \vec{V} \, d^3x = \int_S \vec{V} \cdot \hat{n} \, dS. \tag{9.2}$$

Let's examine this formula. On the left side we have a volume integral over the interior of the blob. The integrand is the scalar function defined by the divergence of $\vec{V}$.

On the right sides is also an integral, taken over the outer surface of the blob S. We can think of it as a sum over all the little surface elements that make up S. Each little surface element has an outward-pointing unit vector $\hat{n}$, and the integrand of the surface integral is the dot product of $\vec{V}$ with $\hat{n}$. Another way to say this is that the integrand is the normal (perpendicular to S) component of $\vec{V}$.

An important special case is a vector field $\vec{V}$ that is spherically symmetric. This means two things: First, it means that the field everywhere points along the radial direction. In addition, spherical symmetry means that the magnitude of $\vec{V}$ depends only on the distance from the origin and not the angular location. If we define a unit vector $\hat{r}$ that points outward from the origin at each point of space, then a spherically symmetric field has the form

$$\vec{V} = V(r)\hat{r},$$

where $V(r)$ is some function of distance from the origin. Consider a sphere of radius r, centered at the origin:

$$x^2 + y^2 + z^2 = r^2.$$

The term *sphere* refers to this two-dimensional shell. The volume contained within the shell is called a *ball*. Now let's apply Gauss's theorem by integrating $\vec{\nabla} \cdot \vec{V}$ over the ball. The left side of Eq. 9.2 becomes the volume integral

$$\int_B \vec{\nabla} \cdot \vec{V} \, d^3x,$$

where B now means the ball. The right side is an integral over the *boundary* of the ball—in other words, over the sphere. Because the field is spherically symmetric, $V(r)$ is constant on the spherical boundary, and the calculation is easy. The integral

$$\int_S \vec{V} \cdot \hat{n} \, dS$$

becomes

$$\int_S V(r)\hat{r} \cdot \hat{r} \, dS,$$

or

$$V(r) \int_S dS.$$

But the integral of dS over the surface of a sphere is just the area of that sphere, $4\pi r^2$. The net result is that Gauss's theorem takes the form

$$\int_B \vec{\nabla} \cdot \vec{V} d^3 x = 4\pi r^2 V(r) \qquad (9.3)$$

for a spherically symmetrical field $\vec{V}$.

9.1.2 Stokes's Theorem

Stokes's theorem also relates an integral over a region to an integral over its boundary. This time the region is not a three-dimensional volume but a two-dimensional surface S bounded by a curve C. Imagine a closed curve in space formed by a thin wire. The two-dimensional surface is like a soap film attached to the wire. Fig. 9.2 depicts such a bounded surface as a shaded region bounded by a curve.

It is also important to give the surface a sense of orientation by equipping it with a unit normal vector, $\hat{n}$. At every point on the surface we imagine a vector $\hat{n}$ that distinguishes one side of the surface from the other.

The left side of Stokes's theorem is an integral over the shaded surface. This integral involves the curl of $\vec{V}$; more precisely, the integrand is the component of $\vec{\nabla} \times \vec{V}$ in the direction $\hat{n}$. We write the integral as

$$\int_S (\vec{\nabla} \times \vec{V}) \cdot \hat{n} dS.$$

Let's examine this integral. We imagine dividing the surface

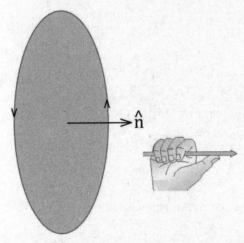

Figure 9.2: Stokes's Theorem and the Right-Hand Rule. The gray surface is not necessarily flat. It can balloon out to the left or right like a soap bubble or a rubber membrane.

S into infinitesimal surface elements dS. At every point, we construct the curl of $\vec{V}$ and take its dot product with the unit normal vector $\hat{n}$ at that point. We multiply the result by the area element dS and add them all up, thus defining the surface integral $\int_S (\vec{\nabla} \times \vec{V}) \cdot \hat{n} \, dS$. That's the left side of Stokes's theorem.

The right side involves an integral over the curved boundary line of S. We need to define a sense of orientation along the curve, and that's where the so-called right-hand rule comes in. This mathematical rule does not depend on human physiology. Nevertheless, the easiest way to explain it is with your right hand. Point your thumb along the vector $\hat{n}$. Your fingers will wrap around the bounding curve C in a particular direction. That direction defines an orientation for the curve C. That's called the right-hand rule.

We may think of the curve as a collection of infinitesimal vectors $\vec{dl}$, each pointing along the curve in the direction specified by the right-hand rule. Stokes's theorem then relates the line integral

$$\oint_C \vec{V} \cdot \vec{dl}$$

around the curve C to the surface integral

$$\int_S \left(\vec{\nabla} \times \vec{V} \right) \cdot \hat{n} \ dS.$$

In other words, Stokes's theorem tells us that

$$\int_S \left(\vec{\nabla} \times \vec{V} \right) \cdot \hat{n} \ dS = \oint_C \vec{V} \cdot \vec{dl}. \tag{9.4}$$

9.1.3 Theorem Without a Name

Later we will need a theorem that to my knowledge has no name. There are lots of theorems that have no name—here's a famous one:

$$1 + 1 = 2. \tag{9.5}$$

There are surely many more theorems without names than with names, and what follows is one such. First, we need a notation. Let $F(x, y, z)$ be a scalar function of space. The gradient of F is a field $\vec{\nabla} F$ with three components

$$\frac{\partial F}{\partial x}, \frac{\partial F}{\partial y}, \frac{\partial F}{\partial z}.$$

Now consider the divergence of $\vec{\nabla} F$. Call it $\vec{\nabla} \cdot \vec{\nabla} F$, or more simply, $\nabla^2 F$. Here it is expressed in concrete form,

$$\nabla^2 F = \frac{\partial^2 F}{\partial x^2} + \frac{\partial^2 F}{\partial y^2} + \frac{\partial^2 F}{\partial z^2}. \tag{9.6}$$

The symbol ∇^2 is called the Laplacian after the French mathematician Pierre-Simon Laplace. The Laplacian stands for the sum of second derivatives with respect to x, y, and z.

The new theorem involves a vector field $\vec{V}$. It uses a vector version of the Laplacian, described in Appendix B. Begin by constructing the curl, $\vec{\nabla} \times \vec{V}$, which is *also* a vector field. As such, we can also take *its* curl,

$$\vec{\nabla} \times (\vec{\nabla} \times \vec{V}).$$

What kind of quantity is this double curl? Taking the curl of any vector field gives another vector field, and so $\vec{\nabla} \times (\vec{\nabla} \times \vec{V})$ is also a vector field. I will now state the un-named theorem:[2]

$$\vec{\nabla} \times (\vec{\nabla} \times \vec{V}) = \vec{\nabla}(\vec{\nabla} \cdot \vec{V}) - \vec{\nabla}^2 \vec{V}. \tag{9.7}$$

Let me say it in words: The curl of the curl of $\vec{V}$ equals the gradient of the divergence of $\vec{V}$ minus the Laplacian of $\vec{V}$. Again, I'll repeat it with some parentheses:

The curl of (the curl of $\vec{V}$) equals the gradient of (the divergence of $\vec{V}$) minus the Laplacian of $\vec{V}$.

How can we prove Eq. 9.7? In the most boring way possible: write out all the terms explicitly and compare both sides. I'll leave it as an exercise for you to work out.

Later on in the lecture, we will use the un-named theorem. Fortunately, we will only need the special case where the divergence of $\vec{V}$ is zero. In that case Eq. 9.7 takes the simpler form,

$$\vec{\nabla} \times (\vec{\nabla} \times \vec{V}) = -\vec{\nabla}^2 \vec{V}. \tag{9.8}$$

---

[2] One of our reviewers suggested calling it the "double-cross" theorem.

9.2 Laws of Electrodynamics

Art: *Ouch, ouch! Gauss, Stokes, divergence, curl—Lenny, Lenny, my head is exploding!*

Lenny: *Oh, that's great, Art. It'll give us a good example of Gauss's theorem. Let's describe the gray matter by a density ρ and a current $\vec{j}$. As your head explodes, the conservation of gray matter will be described by a continuity equation, right? Art? Art? Are you okay?*

9.2.1 The Conservation of Electric Charge

At the heart of electrodynamics is the law of local conservation of electric charge. In Lecture 8, I explained that this means the density and current of charge satisfy the continuity equation (Eq. 8.23),

$$\dot{\rho} = -\vec{\nabla} \cdot \vec{j}.$$

The continuity equation is a local equation that holds at every point of space and time; it says much more than "the total charge never changes." It implies that if the charge changes—let's say in the room where I'm giving this lecture—that it can only do so by passing through the walls of the room. Let's expand on this point.

Take the continuity equation and integrate it over a blob-like region of space, B, bounded by a surface S. The left side becomes

$$\frac{d}{dt} \int_B \rho \, d^3x = \frac{dQ}{dt},$$

where Q is the amount of charge in the region B. In other words the left side is the rate of change of the amount of charge in B. The right side is

$$-\int_B \vec{\nabla} \cdot \vec{j} d^3 x.$$

Here is where we get to use Gauss's theorem. It says

$$\int_B \vec{\nabla} \cdot \vec{j} d^3 x = \int_S \vec{j} \cdot \hat{n} \ dS.$$

Now recall that $\vec{j} \cdot \hat{n}$ is the rate at which charge is crossing the surface per unit time per unit area. When it's integrated, it gives the total amount of charge crossing the boundary surface per unit time. The integrated form of the continuity equation becomes

$$\frac{dQ}{dt} = -\int_S \vec{j} \cdot \hat{n} \ dS, \tag{9.9}$$

and it says that the change in the charge inside B is accounted for by the flow of charge through the surface S. If the charge inside B is decreasing, that charge must be flowing through the boundary of B.

9.2.2   From Maxwell's Equations to the Laws of Electrodynamics

Maxwell's equations were not derived by Maxwell from principles of relativity, least action, and gauge invariance. They were derived from the results of experiment. Maxwell was fortunate; he didn't have to do the experiments himself. Many people, the Founding Fathers, had contributed—Franklin, Coulomb, Ørsted, Ampère, Faraday, and others. Basic principles had been codified by Maxwell's time, especially by Faraday. Most courses in electricity and magnetism take the historical path from the Founding Fathers to

Maxwell. But as in many cases, the historical route is the least clear, logically. What I will do in this lecture is go backward, from Maxwell to Coulomb, Faraday, Ampère, and Ørsted, and then finish off with Maxwell's crowning achievement.

At the time of Maxwell several constants appeared in the theory of electromagnetism—more constants than were strictly necessary. Two of them were called ϵ_0 and μ_0. For the most part, only the product $\epsilon_0\mu_0$ ever appears in equations. In fact, that product is nothing but the constant $1/c^2$, the same c that appears in Einstein's equation $E = mc^2$. Of course, when the founders did their work, these constants were not known to have anything to do with the speed of light. They were just constants deduced from experiments on charges, currents, and forces. So let's forget that c is the speed of light and just take it to be a constant in Maxwell's distillation of the laws laid down by earlier physicists. Here are the equations from Table 8.1, modified to include the appropriate factors of c.[3]

$$\vec{\nabla} \cdot \vec{B} = 0$$

$$\vec{\nabla} \times \vec{E} + \frac{\partial B}{\partial t} = 0$$

$$\vec{\nabla} \cdot \vec{E} = \rho$$

$$c^2 \vec{\nabla} \times \vec{B} - \frac{\partial \vec{E}}{\partial t} = \vec{j} \qquad (9.10)$$

---

[3] In the SI units adopted by many textbooks, you'll see ρ/ϵ_0 and $\vec{j}/\epsilon_0$ instead of ρ and $\vec{j}$.

9.2.3   Coulomb's Law

Coulomb's law is the electromagnetic analog of Newton's gravitational force law. It states that between any two particles there is a force proportional to the product of their electric charges and inversely proportional to the square of the distance between them. Coulomb's law is the usual starting point for a course on electrodynamics, but here we are, many pages into *Volume III*, and we've hardly mentioned it. That's because we are going to derive it, not postulate it.

Let's imagine a point charge Q at the origin, $x = y = z = 0$. The charge density is concentrated at a point and we can describe it as a three-dimensional delta function,

$$\rho = Q\delta^3(x). \tag{9.11}$$

We used the delta function in Lecture 5 to represent a point charge. As in Lecture 5, the symbol $\delta^3(x)$ is shorthand for $\delta(x)\delta(y)\delta(z)$. Because of the delta function's special properties, if we integrate ρ over space we get the total charge Q.

Now let's look at the third Maxwell equation from Eqs. 9.10,

$$\vec{\nabla} \cdot \vec{E} = \rho.$$

What happens to the left side if we integrate this equation over a sphere of radius r? From the symmetry of the situation we expect the field of the point charge to be spherically symmetric, and therefore we can use Eq. 9.3. Plugging in $\vec{E}$, the left side of Eq. 9.3 becomes

$$\int \vec{\nabla} \cdot \vec{E} d^3x.$$

But the third Maxwell equation, $\vec{\nabla} \cdot \vec{E} = \rho$, tells us we can rewrite this as

$$\int \rho d^3 x,$$

which is, of course, the charge Q. The right side of Eq. 9.3 is

$$4\pi r^2 E(r),$$

where $E(r)$ is the electric field at distance r from the charge. Thus Eq. 9.3 becomes

$$Q = 4\pi r^2 E(r).$$

Or, to put it another way, the radial electric field produced by a point charge Q at the origin is given by

$$E(r) = \frac{Q}{4\pi r^2}. \tag{9.12}$$

Now consider a second charge q at a distance r from Q. From the Lorentz force law (Eq. 6.1), the charge q experiences a force $q\vec{E}$ due to the field of Q. The magnitude of the force on q is

$$F = \frac{qQ}{4\pi r^2}. \tag{9.13}$$

This is, of course, the Coulomb law of forces between two charges. We have derived it, not assumed it.

9.2.4 Faraday's Law

Let's consider the work done on a charged particle when moving it from point a to point b in an electric field. The work done in moving a particle an infinitesimal distance $\vec{dl}$ is

$$dW = \vec{F} \cdot \vec{dl},$$

where F is the force on the particle. To move the particle from a to b, the force F does work

$$\int_a^b \vec{F} \cdot \vec{dl}.$$

If the force is due to an electric field, then the work is

$$W = q \int_a^b \vec{E} \cdot \vec{dl}. \tag{9.14}$$

In general, the work depends not only on the end points but also on the path along which the particle is transported. The work done can be nonzero even if the path is a closed loop in space that starts and ends at point a. We can describe this situation by a line integral around a closed loop in space,

$$W = q \oint_C \vec{E} \cdot \vec{dl}. \tag{9.15}$$

Can we actually do work on a particle by moving it around a closed path—for example, by moving it through a closed loop of conducting wire? Under certain circumstances we can. The integral $\oint_C \vec{E} \cdot \vec{dl}$ is called the electromotive force (EMF) for a circuit made from a wire loop.

Let's explore this EMF by using Stokes's theorem (Eq. 9.4). Applied to the electric field, it says

$$\int \left(\vec{\nabla} \times \vec{E} \right) \cdot \hat{n} \, dS = \oint_C \vec{E} \cdot \vec{dl}. \tag{9.16}$$

This allows us to write the EMF for a closed loop of wire in the form

$$EMF = \int \left(\vec{\nabla} \times \vec{E} \right) \cdot \hat{n} \, dS, \tag{9.17}$$

where the integral is over any surface that has the closed curve C as a boundary.

Now comes the point where Maxwell's equations have something to say. Recall the second Maxwell equation from Eqs. 9.10:

$$\vec{\nabla} \times \vec{E} = -\frac{\partial B}{\partial t}.$$

If the magnetic field does not vary with time, then the curl of the electric field is zero, and from Eq. 9.17 the EMF for a closed path is also zero. In situations where the fields do not vary with time, no work is done when carrying a charge around a closed path. This is often the framework for elementary discussions.

But magnetic fields sometimes do vary with time; for example, just wave a magnet around in space. Applying the second Maxwell equation to Eq. 9.17,

$$EMF = -\int \frac{\partial \vec{B}}{\partial t} \cdot \hat{n} \; dS$$

or

$$EMF = -\frac{d}{dt} \int \vec{B} \cdot \hat{n} \; dS.$$

Thus, the EMF is the negative of the time rate of change of a certain quantity $\int \vec{B} \cdot \hat{n} \; dS$. This quantity, the integral of the magnetic field over the surface bounded by the closed wire, is called the magnetic flux through the circuit and is denoted Φ. Our equation for the EMF can be succinctly written as

$$EMF = -\frac{d\Phi}{dt}. \tag{9.18}$$

The EMF represents a force (per unit charge) that pushes a charged particle around the closed loop of wire. If the wire is

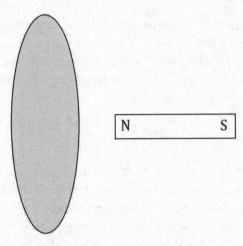

Figure 9.3: Faraday's Law.

an electric conductor, it will cause a current to flow in the wire.

The remarkable fact that an EMF in a circuit can be generated by varying the flux through the circuit was discovered by Michael Faraday and is called Faraday's law. In Fig. 9.3, a closed loop of wire is shown next to a bar magnet. By moving the magnet closer to and farther from the loop, the flux through the loop can be varied and an EMF generated in the loop. The electric field producing the EMF drives a current through the wire. Pull the magnet away from the wire loop and a current flows one way. Push the magnet closer to the loop and the current flows the other way. That's how Faraday discovered the effect.

9.2.5   Ampère's Law

How about the other Maxwell equations? Can we recognize any other basic electromagnetic laws by integrating them?

Let's try the fourth Maxwell equation from Eqs. 9.10,

$$c^2 \vec{\nabla} \times \vec{B} - \frac{\partial \vec{E}}{\partial t} = \vec{j}.$$

For now, we'll assume nothing varies with time; the current $\vec{j}$ is steady and all the fields are static. In that case, the equation takes the simpler form

$$\vec{\nabla} \times \vec{B} = \frac{\vec{j}}{c^2}. \tag{9.19}$$

One thing we see from this equation is that the magnetic field will be very small unless a huge current flows through the wire. In ordinary laboratory units the factor $1/c^2$ is a very small number. That fact played an important role in the interlude on crazy units that followed Lecture 5.

Let's consider a current flowing in a very long, thin wire— so long that we may think of it as infinite. We can orient our axes so that the wire lies on the x axis and the current flows to the right. Since the current is confined to the thin wire, we may express it as a delta function in the y and z directions:

$$j_x = j\delta(y)\delta(z)$$
$$j_y = 0$$
$$j_z = 0. \tag{9.20}$$

Now imagine a circle of radius r surrounding the wire, as shown in Fig. 9.4. This time the circle is an imaginary mathematical circle, not a physical wire. We are going to use Stokes's theorem again, in a way that should be clear from the figure. What we'll do is integrate Eq. 9.19 over the disclike shaded region in Fig. 9.4. The left side gives the integral of $\vec{\nabla} \times \vec{B}$, which by Stokes's theorem is the line integral of B over the circle. The right side just gives $1/c^2$ times the numerical value of the current through the wire, j.

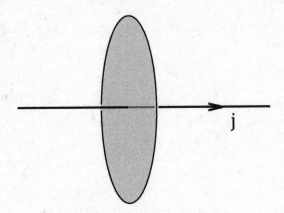

Figure 9.4: Ampère's Law.

Thus,

$$\oint \vec{B} \cdot \vec{dl} = \frac{j}{c^2}. \qquad (9.21)$$

It follows that a current through a wire produces a magnetic field that circulates around the wire. By that I mean that the field points along the angular direction, not along the x axis or in the direction away from the wire.

Because $\vec{B}$ is parallel to $\vec{dl}$ at every point along the loop, the integral of the magnetic field is just the magnitude of the field at distance r times the circumference of the circle, giving

$$2\pi r B(r) = \frac{j}{c^2}.$$

Solving for $B(r)$, we find that the magnetic field produced by a current-carrying wire at a distance r from the wire has the value

$$B(r) = \frac{j}{2\pi r c^2}. \qquad (9.22)$$

On the face of it, one might think the factor $1/c^2$ would be so overwhelmingly small that no magnetic field would ever be detectable from an ordinary current. Indeed, in the usual metric system,

$$\frac{1}{c^2} \approx 10^{-17}.$$

What works to compensate for this tiny number is the huge number of electrons that move through the wire when an EMF is created.

Equation 9.22 is called Ørsted's law after the Danish physicist Hans Christian Ørsted. Ørsted noted that an electric current in a wire would cause the magnetized needle of a compass to align along a direction perpendicular to the wire and along the angular direction around the wire. Within a short time the French physicist André-Marie Ampère generalized Ørsted's law to more general flows of current. The Ørsted-Ampère laws together with Faraday's law were the basis for Maxwell's investigations that led to his equations.

9.2.6 Maxwell's Law

By Maxwell's law I mean the fact (discovered by James Clerk Maxwell in 1862) that light is composed of electromagnetic waves—wavelike undulations of electric and magnetic fields. The discovery was a mathematical one, not based on a laboratory experiment. It consisted of showing that Maxwell's equations have wavelike solutions that propagate with a certain speed, which Maxwell calculated. By that time, the speed of light had been known for almost two hundred years, and Maxwell observed that the speed of his electromagnetic waves agreed. What I will do in the rest of this lecture is show that Maxwell's equations imply that electric and magnetic fields satisfy the wave equation.

We are going to consider Maxwell's equations in regions of space where there are no sources—in other words, no currents or charge densities:

$$\vec{\nabla} \cdot \vec{B} = 0$$

$$\vec{\nabla} \cdot \vec{E} = 0$$

$$\vec{\nabla} \times \vec{E} + \frac{\partial \vec{B}}{\partial t} = 0$$

$$c^2 \vec{\nabla} \times \vec{B} - \frac{\partial \vec{E}}{\partial t} = 0. \tag{9.23}$$

Our goal is to show that the electric and magnetic fields satisfy the same kind of wave equation that we studied in Lecture 4. Let's grab the simplified wave equation (Eq. 4.26) and transpose all the space derivatives to the right side:

$$\frac{1}{c^2} \frac{\partial^2 \phi}{\partial t^2} = \frac{\partial^2 \phi}{\partial x^2} + \frac{\partial^2 \phi}{\partial y^2} + \frac{\partial^2 \phi}{\partial z^2}.$$

The Laplacian provides a convenient shorthand for the entire right side; we can rewrite the wave equation as

$$\frac{1}{c^2} \frac{\partial^2 \phi}{\partial t^2} = \nabla^2 \phi. \tag{9.24}$$

As I explained in Lecture 4, this equation describes waves that move with the velocity c.

Now let's put ourselves in the shoes of Maxwell. Maxwell had his equations (Eqs. 9.23), but those equations were not at all similar to Eq. 9.24. In fact, he had no obvious reason to connect the constant c with any velocity, let alone the velocity of light. I suspect the question he asked was this:

I have some coupled equations for $\vec{E}$ and $\vec{B}$. Is there some way to simplify things by eliminating one of them (either $\vec{E}$ or $\vec{B}$) to obtain equations just for the other?

Of course I wasn't there, but it's easy to imagine Maxwell asking this question. Let's see if we can help him. Take the last equation from Eqs. 9.23 and differentiate both sides with respect to time. We get

$$c^2 \vec{\nabla} \times \frac{\partial \vec{B}}{\partial t} - \frac{\partial^2 \vec{E}}{\partial t^2} = 0.$$

Now let's use the third Maxwell equation to replace $\dfrac{\partial \vec{B}}{\partial t}$ with $-\vec{\nabla} \times \vec{E}$, which indeed gives an equation for the electric field,

$$\frac{\partial^2 \vec{E}}{\partial t^2} + \vec{\nabla} \times \left(\vec{\nabla} \times \vec{E} \right) = 0.$$

This is beginning to look more like a wave equation, but we are not quite there. The last step is to use the unnamed theorem (Eq. 9.7). Fortunately, all we need is the simplified form, Eq. 9.8. The reason is that the second Maxwell equation, $\vec{\nabla} \cdot \vec{E} = 0$, is exactly the condition that allows the simplification. The result is just what we want:

$$\frac{1}{c^2} \frac{\partial^2 \vec{E}}{\partial t^2} = \vec{\nabla}^2 \vec{E}. \tag{9.25}$$

Each component of the electric field satisfies the wave equation.

Lenny: *I wasn't there Art, but I can well imagine Maxwell's excitement. I can hear him saying to himself:*

What are these waves? From the form of the equation, the wave velocity is that funny constant c that appears all over my equations. If I'm not mistaken, in metric units it is about 3×10^8. *YES!* 3×10^8 *meters per second! The speed of light!*

And thus it came to pass that James Clerk Maxwell discovered the electromagnetic nature of light.

Lecture 10

Maxwell From Lagrange

Art: *I'm worried about Maxwell. He's been talking to himself for a long time.*

Lenny: *Don't worry, Art, it looks like he's close to a solution.* [Lenny gestures toward the front door] *Besides, help has just arrived.*

Two gentlemen walk into the Hideaway. One of them is elderly and nearly blind but has no trouble finding his way. They take seats on either side of Maxwell. Maxwell recognizes them immediately.

Maxwell: *Euler! Lagrange! I was hoping you'd show up! Your timing is perfect!*

In this lecture, we'll do two things. The first is to follow up on Lecture 9 and work out the details of electromagnetic plane waves. Then, in the second part of the lecture, we'll introduce the action principle for the electromagnetic field and derive the two Maxwell equations that are not identities. To help keep track of where we've been and where we're going, I've summarized the full set of Maxwell equations in Table 10.1.

Form	Equation
3-vector:	$\vec{\nabla} \cdot \vec{B} = 0$
	$\vec{\nabla} \times \vec{E} + \dfrac{\partial}{\partial t} \vec{B} = 0$
4-vector:	$\partial_\mu F_{\nu\sigma} + \partial_\nu F_{\sigma\mu} + \partial_\sigma F_{\mu\nu} = 0$
3-vector:	$\vec{\nabla} \cdot \vec{E} = \rho$
	$\vec{\nabla} \times \vec{B} - \dfrac{\partial}{\partial t} \vec{E} = \vec{j}$
4-vector:	$\partial_\nu F^{\mu\nu} = J^\mu$

Table 10.1: Maxwell equations in 3-vector and 4-vector form. The top group is derived from identities; the bottom group is derived from the action principle.

10.1 Electromagnetic Waves

In Lectures 4 and 5 we discussed waves and wave equations, and in Lecture 9 we saw how Maxwell derived the wave equation for the components of the electric and magnetic fields.

Let's begin with the Maxwell equations (Eqs. 9.23) in the absence of sources. I'll write them here for convenience:

$$\vec{\nabla} \cdot \vec{B} = 0$$

$$\vec{\nabla} \cdot \vec{E} = 0$$

$$\vec{\nabla} \times \vec{E} + \frac{\partial \vec{B}}{\partial t} = 0$$

$$c^2 \vec{\nabla} \times \vec{B} - \frac{\partial \vec{E}}{\partial t} = 0.$$

Now let's consider a wave moving along the z axis with wavelength

$$\lambda = \frac{2\pi}{k},$$

where k is the so-called wave number. We can choose any wavelength we like. A generic plane wave has the functional form

$$Field\ Value = C \sin(kz - \omega t),$$

where C is any constant. When applied to the electric field, the components have the form

$$E_x(t, z) = \mathcal{E}_x \sin(kz - \omega t) \tag{10.1}$$

$$E_y(t, z) = \mathcal{E}_y \sin(kz - \omega t) \tag{10.2}$$

$$E_z(t, z) = \mathcal{E}_z \sin(kz - \omega t), \tag{10.3}$$

where $\mathcal{E}_x$, $\mathcal{E}_y$ and $\mathcal{E}_z$ are numerical constants. We can think of them as the components of a fixed vector that defines the *polarization* direction of the wave.

We've assumed that the wave propagates along the z axis. Obviously this is not very general; the wave could propagate in any direction. But we are always free to realign our axes so that the motion of the wave is along z.

Another freedom we have is to add a cosine dependence; but this simply shifts the wave along the z axis. By shifting the origin we can get rid of the cosine.

Now let's use the equation $\vec{\nabla} \cdot \vec{E} = 0$. Because z is the only space coordinate that the electric field components depend on, this equation takes an especially simple form,

$$\frac{\partial E_z}{\partial z} = 0. \tag{10.4}$$

If we combine this with Eq. 10.3, we find that

$$\mathcal{E}_z = 0.$$

In other words, the component of the electric field along the direction of propagation must be zero. Waves with this property are called *transverse waves*.

Lastly, we can always align the x and y axes so that the polarization vector $\vec{\mathcal{E}}$ lies along the x axis. Thus the electric field has the form,

$$E_z = 0$$

$$E_y = 0$$

$$E_x = \mathcal{E}_x \sin{(kz - \omega t)}. \tag{10.5}$$

Next, let's consider the magnetic field. We might try allowing the magnetic field to propagate in a different direction from the electric field, but that would violate the Maxwell equations that involve both the electric and magnetic fields. The magnetic field must also propagate along the z axis and have the form

$$B_x(t, z) = \mathcal{B}_x \sin(kz - \omega t) \tag{10.6}$$

$$B_y(t, z) = \mathcal{B}_y \sin(kz - \omega t) \tag{10.7}$$

$$B_z(t, z) = \mathcal{B}_z \sin(kz - \omega t). \tag{10.8}$$

By the same argument we used for the electric field, the equation $\vec{\nabla} \cdot \vec{B} = 0$ implies that the z component of the magnetic field is zero. Therefore the magnetic field must also lie in the x, y plane, but not necessarily in the same direction as the electric field. In fact, it must be perpendicular to the electric field and therefore must lie along the y axis. To see this property we use the Maxwell equation

$$\vec{\nabla} \times \vec{E} + \frac{\partial \vec{B}}{\partial t} = 0.$$

In component form, this becomes

$$\dot{B}_x = -\left(\frac{\partial E_z}{\partial y} - \frac{\partial E_y}{\partial z} \right)$$

$$\dot{B}_y = -\left(\frac{\partial E_x}{\partial z} - \frac{\partial E_z}{\partial x} \right)$$

$$\dot{B}_z = -\left(\frac{\partial E_y}{\partial x} - \frac{\partial E_x}{\partial y} \right). \tag{10.9}$$

Now, keeping in mind that E_z and E_y are zero, and that the fields only vary with respect to z, we see that only the y component of the magnetic field can vary with time. Given that we are discussing oscillating waves, it follows that only the y component of $\vec{B}$ can be nonzero.

One more fact follows from Eqs. 10.9. If we plug in the forms $E_x = \mathcal{E}_x \sin(kz - \omega t)$ and $B_y = \mathcal{B}_y \sin(kz - \omega t)$, we will find that $\mathcal{B}_y$ is constrained to be

$$\mathcal{B}_y = \frac{k}{\omega}\mathcal{E}_x. \tag{10.10}$$

There is still one more Maxwell equation we haven't used, namely

$$c^2 \vec{\nabla} \times \vec{B} - \frac{\partial \vec{E}}{\partial t} = 0.$$

If we take the x component of this last Maxwell equation and use what we have already learned, we find that it reduces to

$$\frac{1}{c^2}\frac{\partial E_x}{\partial t} = -\frac{\partial B_y}{\partial z}.$$

Plugging in the forms of E_x and B_y, out comes a simple relation between the frequency ω and the wave number k:

$$\omega = ck.$$

The wave form $\sin(kz - \omega t)$ becomes

$$\sin k(z - ct). \tag{10.11}$$

This is exactly the form of a wave propagating along the z axis with velocity c. Let's summarize the properties of electromagnetic plane waves:

- They propagate along a single axis at the speed of light.
- Electromagnetic waves are transverse, meaning that the fields lie in the plane perpendicular to the propagation axis.

- The electric and magnetic fields are perpendicular to one another.
- The ratio of the electric and magnetic fields is

$$\frac{\mathcal{B}_y}{\mathcal{E}_x} = \frac{1}{c}.$$

In relativistic units (where $c = 1$), the magnitudes of the electric and magnetic fields are equal.

I'll remind you of one property of light waves that you're probably familiar with, especially if you buy a better quality of sunglasses. Light is polarized. In fact, all electromagnetic waves are polarized. The direction of polarization is the direction of the electric field, so in our example we would say that the wave is polarized along the x axis. The properties of plane electromagnetic waves can be visualized in a single picture, shown in Fig. 10.1.

10.2 Lagrangian Formulation of Electrodynamics

I've made this point so many times that I risk being repetitive. Once again: The fundamental ideas of energy conservation, momentum conservation, and the relationship between conservation laws and symmetry laws follow *only* if you start with the principle of least action. You can write all the differential equations you like, and they may be mathematically consistent. All the same, there will be no energy conservation—*there will be no energy to be conserved*—unless those equations are derived from a Lagrangian and an action principle. Since we're inclined to deeply believe in energy conservation, we should look for a Lagrangian formulation of Maxwell's equations.

Let's go through the fundamental principles and make our

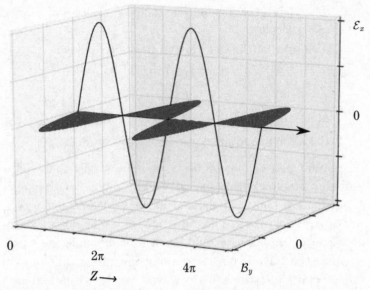

Figure 10.1: Snapshot of an Electromagnetic Plane Wave Propagating to the Right and Out of the Page (positive z direction). The $\vec{E}$ and $\vec{B}$ fields are perpendicular to each other, and to the direction of propagation. The $\vec{B}$ field is shaded.

best guess about the Lagrangian. Remember that the two equations

$$\vec{\nabla} \cdot \vec{B} = 0$$

$$\vec{\nabla} \times \vec{E} + \frac{\partial \vec{B}}{\partial t} = 0$$

are mathematical identities that follow from the definitions of $\vec{E}$ and $\vec{B}$ (in terms of the vector potential). There is no more need to derive them from an action principle than there is need to derive $1 + 1 = 2$.

Our goal now is to derive the second set. In 3-vector

notation, these equations are

$$\vec{\nabla} \cdot \vec{E} = \rho \qquad (10.12)$$

$$\vec{\nabla} \times \vec{B} - \frac{\partial \vec{E}}{\partial t} = \vec{j}. \qquad (10.13)$$

We've identified the charge density ρ as the time component of the current 4-vector J^ν,

$$\rho = J^0.$$

Likewise, the three components of $\vec{j}$ correspond to the three space components of J^ν,

$$\vec{j} = (J^1, J^2, J^3)$$

or

$$\vec{j} = J^m.$$

All these relationships are captured in covariant form by the single equation

$$\partial_\mu F^{\mu\nu} = -J^\nu. \qquad (10.14)$$

The time component of this equation is equivalent to Eq. 10.12, and the three space components conspire to give us Eq. 10.13. How do we derive these equations from a Lagrangian? Before we can even try, we must guess what the correct Lagrangian is. The fundamental principles will help narrow the search.

We've already discussed the need for an action principle. We have also seen that the mathematics of the action principle plays out slightly differently for fields than it does for particles. Here we focus on fields because the Maxwell equations are field equations. Let's see what the other principles can teach us.

10.2.1   Locality

Whatever happens at a particular time and position can only be related to things that happen at neighboring times and positions. How do we guarantee that? We make sure the Lagrangian density inside the action integral depends only on the fields themselves and on their first derivatives with respect to the coordinates X^μ.

In general, the Lagrangian density depends on each field in the theory; there could be several of them, but I'll just use the single variable ϕ to represent all the fields that exist.[1] The Lagrangian also depends on the derivatives of ϕ with respect to space and time. We're already accustomed to the notation $\partial_\mu \phi$. We'll be using that, along with an even more condensed notation, $\phi_{,\mu}$, which means the same thing:

$$\frac{\partial \phi}{X^\mu} = \partial_\mu \phi = \phi_{,\mu}.$$

This notation is standard. In the symbol $\phi_{,\mu}$ the comma means *derivative*, and μ means *with respect to* X^μ. The locality requirement says that the Lagrangian density depends only on ϕ, and $\phi_{,\mu}$. In other words, the action integral must have the form

$$Action = \int d^4x \mathcal{L}(\phi, \phi_{,\mu}).$$

The symbol $\phi_{,\mu}$ doesn't mean one *particular* derivative. It's a generic symbol that refers to derivatives with respect to all components, time and space. By requiring the action integral to take this form, we guarantee that the resulting equations of motion will be differential equations that relate things locally.

---

[1] For electromagnetism, the fields turn out to be components of the vector potential, but let's not get ahead of ourselves.

10.2.2 Lorentz Invariance

Lorentz invariance is a simple requirement: The Lagrangian density needs to be a scalar. It must have the same value in every reference frame. We could just leave it at that and move on to the next principle. Instead, I'd like to provide some context by quickly reviewing our previous results for scalar fields. The Maxwell equations are based on vector fields, not on scalar fields, and it's good to see how they compare.

For a scalar field ϕ, the Lagrangian may contain ϕ itself or any function of ϕ. Let's call this function $U(\phi)$. The Lagrangian can also contain derivatives.[2] On the other hand, there are some things it cannot contain. For example, it can't contain $\partial_x \phi$ all by itself. That would be nonsense because $\partial_x \phi$ is not a scalar. It's the x component of a vector. While we can't just throw in arbitrary components of a vector, we're allowed to carefully package them up to form a scalar. The quantity

$$\partial_\mu \phi \partial^\mu \phi = \phi_{,\mu} \phi^{,\mu}$$

is a perfectly good scalar and could certainly appear in a Lagrangian. Because μ is a summation index, we can expand this as

$$\partial_\mu \phi \partial^\mu \phi = -(\partial_t \phi)^2 + (\partial_x \phi)^2 + (\partial_y \phi)^2 + (\partial_z \phi)^2,$$

or in "comma" notation,

$$\phi_{,\mu} \phi^{,\mu} = -(\partial_t \phi)^2 + (\partial_x \phi)^2 + (\partial_y \phi)^2 + (\partial_z \phi)^2.$$

In Lecture 4, we used terms like the preceding ones to write a simple Lagrangian (Eq. 4.7). In our newer notation, we can

---

[2] It's not going too far to say that it *must* contain them. Without derivatives, the Lagrangian would be uninteresting.

write that Lagrangian as

$$\mathcal{L} = -\frac{1}{2}\partial_\mu\phi\partial^\mu\phi - U(\phi)$$

or

$$\mathcal{L} = -\frac{1}{2}\phi_{,\mu}\phi^{,\mu} - U(\phi). \qquad (10.15)$$

Based on this Lagrangian, we used the Euler-Lagrange equations to derive the equations of motion. I've written only one field in this example, but there could be several. In general, we might add up the Lagrangians for several fields. We might even combine them in more complicated ways.

For each independent field—in our example there's only one—we begin by taking the partial derivative of the Lagrangian with respect to $\phi_{,\mu}$, which is

$$\frac{\partial\mathcal{L}}{\partial\phi_{,\mu}}.$$

Then we differentiate that term with respect to X^μ to get

$$\frac{\partial}{\partial X^\mu}\frac{\partial\mathcal{L}}{\partial\phi_{,\mu}}.$$

That's the left side of the Euler-Lagrange equation. The right side is just the derivative of $\mathcal{L}$ with respect to ϕ. The complete equation is

$$\frac{\partial}{\partial X^\mu}\frac{\partial\mathcal{L}}{\partial\phi_{,\mu}} = \frac{\partial\mathcal{L}}{\partial\phi}. \qquad (10.16)$$

This is the Euler-Lagrange equation for fields. It's the direct counterpart to the Lagrange equations for particle motion.

As you recall, the equation for particle motion is

$$\frac{d}{dt}\frac{\partial \mathcal{L}}{\partial \dot{q}} = \frac{\partial \mathcal{L}}{\partial q}.$$

To derive the equation of motion, we evaluate the Lagrangian (Eq. 10.15) using the Euler-Lagrange equation (Eq. 10.16). We did this in Section 4.3.3 and found that the equation of motion is

$$\frac{\partial^2 \phi}{\partial t^2} - \frac{\partial^2 \phi}{\partial x^2} - \frac{\partial^2 \phi}{\partial y^2} - \frac{\partial^2 \phi}{\partial z^2} = -\frac{\partial U}{\partial \phi},$$

which is a simple wave equation.

Starting with the vector potential, we will follow exactly the same process to arrive at Maxwell's equations. It's more complicated because there are more indices to keep track of. In the end, I think you'll agree that it's a beautiful piece of work. Before we take that plunge, we have one more principle to review: gauge invariance. This is always a requirement in electrodynamics.

10.2.3 Gauge Invariance

To make sure the Lagrangian is gauge invariant, we construct it from quantities that are themselves gauge invariant. The obvious choices are the components of $F_{\mu\nu}$, the electric and magnetic fields. These fields don't change when you add the derivative of a scalar to A_μ. In other words, they don't change if you make the replacement

$$A_\mu \Longrightarrow A_\mu + \frac{\partial S}{\partial X^\mu}.$$

Therefore, any Lagrangian we construct from the components of F will be gauge invariant.

As we saw in the case of a scalar field, there are also some quantities we *cannot* use. For example, we can't include $A_\mu A^\mu$. This may seem counterintuitive. After all, $A_\mu A^\mu$ is a perfectly good Lorentz invariant scalar.[3] However, it's not gauge invariant. If you add the gradient of a scalar to A, then $A_\mu A^\mu$ would indeed change, so it's no good. It can't be part of the Lagrangian.

10.2.4   The Lagrangian in the Absence of Sources

We'll build up our Lagrangian in two stages. First, we'll consider the case where there's an electromagnetic field but no charges or currents. In other words, the case where the current 4-vector is zero:

$$J^\mu = 0.$$

Later on, we'll modify the Lagrangian to cover nonzero current vectors as well.

We can use the components $F_{\mu\nu}$ in any way we like without worrying about gauge invariance. However, Lorentz invariance requires more care; we must somehow combine them to form a scalar by contracting the indices.

Consider some general tensor $T_{\mu\nu}$ with two indices. It's simple to construct a scalar by raising one index, then contracting. In other words, we could form the expression

$$T_\mu{}^\mu,$$

with μ as a summation index. That's the general technique for making scalars out of tensors. What happens if we try

---

[3] Remember that the μ's in this expression do not signify derivatives. They just represent the components of A. To indicate differentiation, we would use a comma.

that with $F_{\mu\nu}$? It's not hard to see that the result is

$$F_\mu{}^\mu = 0,$$

because all the diagonal components (components with equal indices) are zero. In other words,

$$F_{00} = 0$$

$$F_{11} = 0$$

$$F_{22} = 0$$

$$F_{33} = 0.$$

The expression $F_\mu{}^\mu$ tells us to add these four terms together, and of course the result is zero. This is not a good choice for the Lagrangian. In fact, it appears that any term linear in $F_{\mu\nu}$ would not be a good choice. If linear terms are no good, we can try something that's nonlinear. The simplest nonlinear term would be the quadratic term

$$F_{\mu\nu}F^{\mu\nu}.$$

This expression is nontrivial. It is certainly *not* identically equal to zero. Let's figure out what it means. First, consider the mixed space-and-time components

$$F_{0n}F^{0n}.$$

These are the components where μ is zero and therefore represents time. Latin index n represents the space components of ν. What does the whole expression represent? It's nearly identical to the square of the electric field. As we've seen before, the mixed components of $F_{\mu\nu}$ are the electric field components. The reason $F_{0n}F^{0n}$ is not *exactly* the square of the electric field is that raising one time component produces a change of sign. Because each term contains one upper time index and one lower time index, the result is *minus* the square of $\vec{E}$,

$$F_{0n}F^{0n} = -E^2.$$

A moment's thought shows that this term appears twice in the summation because we can interchange the roles of μ and ν. In other words, we must also consider terms of the form

$$F_{n0}F^{n0} = -E^2.$$

These terms also add up to $-E^2$ because $F_{\mu\nu}$ is antisymmetric. In this second form, the first index represents the space components, while the second index represents time. The overall result of all this squaring and adding is $-2E^2$. In other words,

$$F_{0n}F^{0n} + F_{n0}F^{n0} = -2E^2.$$

You might think that the antisymmetry of $F_{\mu\nu}$ would cause these two sets of terms to cancel. However, they don't because each component is squared.

Now we need to account for the space-space components of $F_{\mu\nu}$. These are the terms where neither index is zero. For example, one such term is

$$F_{12}F^{12}.$$

Raising and lowering the space indices does nothing; there's no change of sign. In terms of electric and magnetic fields, F_{12} is the same as B_3, aka B_z, the z component of magnetic field. Therefore, $F_{12}F^{12}$ is the same as $(B_z)^2$. When we consider its antisymmetric counterpart $F_{21}F^{21}$, we find that the term $(B_z)^2$ enters the sum twice.

We have now accounted for all the terms in the sum $F_{\mu\nu}F^{\mu\nu}$ except the diagonals, where the two indices are equal to each other. But we already know that those components are zero, and so we're done. Combining the space-space terms with the space-time terms results in

$$F_{\mu\nu}F^{\mu\nu} = -2E^2 + 2B^2,$$

or

$$-F_{\mu\nu}F^{\mu\nu} = 2\left(E^2 - B^2\right).$$

By convention, this equation is written in a slightly different form. When I say *by convention*, I mean it would affect nothing if we ignored the convention; the equations of motion would be the same. One convention is that the E^2 term is positive and the B^2 term is negative. The second convention incorporates a factor of one quarter. As a result, the Lagrangian is usually written

$$\mathcal{L} = -\frac{1}{4} F_{\mu\nu}F^{\mu\nu} \tag{10.17}$$

or

$$\mathcal{L} = \frac{1}{2}\left(E^2 - B^2\right). \tag{10.18}$$

The factor of $\frac{1}{4}$ has no physical content. Its only purpose is to maintain consistency with long-standing habits.

10.3 Deriving Maxwell's Equations

Eq. 10.18 is Lorentz invariant, local, and gauge invariant. It's not only the simplest Lagrangian we can write down, it's also the correct one for electrodynamics. In this section, we'll see how it gives rise to the Maxwell equations. The derivation is a little tricky but not as hard as you might think. We'll proceed in small simple steps. To start with, we'll ignore the J^μ terms and work things out for empty space. Then we'll bring J^μ back into the picture.

Once again, here are the Euler-Lagrange equations for fields:

$$\frac{\partial}{\partial X^\nu} \frac{\partial \mathcal{L}}{\partial \phi_{,\nu}} = \frac{\partial \mathcal{L}}{\partial \phi}. \tag{10.19}$$

For each field, we write a separate equation of this kind. What are these fields? They are the vector potential components

$$A_0, A_1, A_2, A_3$$

or

$$A_t, A_x, A_y, A_z.$$

These are four distinct and independent fields. What about their derivatives? To help things along, we'll extend our comma notation to include 4-vectors.[4] In this notation, the symbol $A_{\mu,\nu}$, or "A sub μ comma ν," stands for the derivative of A_μ with respect to X^ν. In other words, we define $A_{\mu,\nu}$ as

$$A_{\mu,\nu} \equiv \frac{\partial A_\mu}{\partial X^\nu}.$$

The comma in the subscript of the left side is essential. It tells you to take the appropriate derivative. Without it the symbol $A_{\mu\nu}$ would not be a derivative at all. It would just be a component of some two-index tensor. What does the field tensor look like in this notation? As we know, the field

---

[4] Until now we've used the comma notation only for derivatives of a scalar.

tensor is defined as

$$F_{\mu\nu} = \frac{\partial A_\nu}{\partial X^\mu} - \frac{\partial A_\mu}{\partial X^\nu}.$$

It's easy to translate this into comma notation. It's just

$$F_{\mu\nu} = A_{\nu,\mu} - A_{\mu,\nu}. \qquad (10.20)$$

Why put so much effort into condensing the notation for partial derivatives? I think you'll see the value of this technique in the next few paragraphs. Of course it saves a lot of writing, but it does much more; it makes the symmetries in our equations pop right out of the page. Even more importantly, it exposes the simplicity of working through the Euler-Lagrange equations.

We'll take the four components of A_μ to be separate independent fields. There's a field equation for each one of them. To translate the Lagrangian (Eq. 10.17) into condensed notation, we simply substitute for $F_{\mu\nu}$ from Eq. 10.20:

$$\mathcal{L} = -\frac{1}{4}\left(A_{\nu,\mu} - A_{\mu,\nu}\right)\left(A^{\nu,\mu} - A^{\mu,\nu}\right). \qquad (10.21)$$

That's the Lagrangian. If you like, you can write it out in detail in terms of derivatives. The components A_μ and A_ν are the things that correspond to ϕ in Eq. 10.19. There's an Euler-Lagrange equation for each component of A. To begin with, we'll choose a single component, A_x, to work on. The first thing to do is compute the partial derivative of $\mathcal{L}$ with respect to $A_{x,\mu}$, that is,

$$\frac{\partial \mathcal{L}}{\partial A_{x,\mu}}.$$

Let's break this calculation into small steps.

We need to calculate the derivative of $\mathcal{L}$ with respect to specific derivatives of specific components of A. To see what that means, consider a case, the derivative of $\mathcal{L}$ with respect to $A_{x,y}$. Eq. 10.21 tells us that $\mathcal{L}$ is a summation over the four values of μ and ν. We could write out all sixteen terms of this expansion and then look for terms that contain $A_{x,y}$. If we did that, we would find that $A_{x,y}$ appears in only two terms that have the form

$$(A_{x,y} - A_{y,x})(A^{x,y} - A^{y,x}).$$

We temporarily ignored the factor of $-\frac{1}{4}$. Because both x and y are space indices, lowering any of the indices makes no difference, and we can rewrite this term more simply as

$$(A_{x,y} - A_{y,x})(A_{x,y} - A_{y,x}).$$

Simplifying further and restoring the numerical factor, this becomes

$$-\frac{1}{2}(A_{x,y} - A_{y,x})^2.$$

Why did I write $\dfrac{1}{2}$ instead of $\dfrac{1}{4}$? Because there are two terms of this form in the expansion; one that occurs when $\mu = x$ and $\nu = y$, and another when $\mu = y$ and $\nu = x$.

At this point, there should be no mystery about how to differentiate $\mathcal{L}$ with respect to $A_{x,y}$. Because none of the other terms contain $A_{x,y}$, they all go to zero when we differentiate and we can ignore them. In other words, we can now write

$$\frac{\partial \mathcal{L}}{\partial A_{x,y}} = \frac{\partial}{\partial A_{x,y}}\left[-\frac{1}{2}(A_{x,y} - A_{y,x})^2\right].$$

This derivative is straightforward as long as we remember that $A_{x,y}$ and $A_{y,x}$ are two different objects. We're only interested in the dependence on $A_{x,y}$, and therefore we regard

$A_{y,x}$ as a constant for this partial derivative. The result is

$$\frac{\partial \mathcal{L}}{\partial A_{x,y}} = -(A_{x,y} - A_{y,x}),$$

and we can now recognize the right side as an element of the field tensor, namely $-F_{yx}$. The result of all this work is

$$\frac{\partial \mathcal{L}}{\partial A_{x,y}} = -F_{yx} = -F^{yx},$$

or, using the antisymmetry of F,

$$\frac{\partial \mathcal{L}}{\partial A_{x,y}} = F_{xy} = F^{xy}.$$

The far right side of this equation, F^{xy}, comes for free because upper indices are equivalent to lower indices for space components.

It took a long time to reach this point, but the result is quite simple. If you go through the same exercise for every other component, you'll discover that each of them follows the same pattern. The general formula is

$$\frac{\partial \mathcal{L}}{\partial A_{\mu,\nu}} = F^{\mu\nu}. \tag{10.22}$$

The next step, according to Eq. 10.19, is to differentiate Eq. 10.22 with respect to X^ν. When we do this to both sides of Eq. 10.22, we get

$$\frac{\partial}{\partial X^\nu} \frac{\partial \mathcal{L}}{\partial A_{\mu,\nu}} = \frac{\partial F^{\mu\nu}}{\partial X^\nu}$$

or

$$\frac{\partial}{\partial X^\nu} \frac{\partial \mathcal{L}}{\partial A_{\mu,\nu}} = \partial_\nu F^{\mu\nu}. \tag{10.23}$$

But wait a minute! The right side of Eq. 10.23 is nothing but the left side of the last Maxwell equation from Table 10.1.

Could it really be that easy? Let's withhold judgment for a moment, until we work out the right side of the Euler-Lagrange equation. That means taking the derivative of $\mathcal{L}$ with respect to the fields themselves. In other words, taking the derivative of $\mathcal{L}$ with respect to the undifferentiated components of A, such as A_x. But the undifferentiated components of A do not even appear in $\mathcal{L}$, and therefore the result is zero. That's all there is to it. The right side of the Euler-Lagrange equation is zero, and the equation of motion for empty space (no charges or currents) matches the Maxwell equation perfectly:

$$\frac{\partial}{\partial X^\nu}\,\frac{\partial \mathcal{L}}{\partial A_{\mu,\nu}} = \frac{\partial F^{\mu\nu}}{\partial X^\nu} = 0.$$

10.4   Lagrangian with Nonzero Current Density

How can we modify the Lagrangian to include J^μ, the current density?[5] We need to add something to $\mathcal{L}$ that includes all the components of J^μ. The current density J^μ has four components: The time component is ρ, and the space components are the three components of $\vec{j}$. The mth space component is the charge per unit area per unit time that passes through a little window oriented along the m axis. In symbols, we can write

$$J^\mu = \rho, j^m.$$

---

[5] Eq. 10.17 is the Lagrangian in terms of $F_{\mu\nu}$. Eq. 10.21 is the same Lagrangian in terms of the vector potential.

If we consider a little boxlike cell in space, the charge inside the cell is ρ times the volume of the cell. In other words, it's ρ times $dx\ dy\ dz$. The rate of change of charge inside the cell is just the time derivative of that quantity. The principle of local charge conservation says that the only way this charge can change is for charges to pass through the walls of the cell. That principle gave us the continuity equation,

$$\frac{\partial \rho}{\partial t} + \vec{\nabla} \cdot \vec{j} = 0, \tag{10.24}$$

which we worked out in Lecture 8. The symbol $\vec{\nabla} \cdot \vec{j}$ (the divergence of $\vec{j}$) is defined as

$$\vec{\nabla} \cdot \vec{j} = \frac{\partial j_x}{\partial x} + \frac{\partial j_y}{\partial y} + \frac{\partial j_z}{\partial z}.$$

The first term on the right side (partial of j_x with respect to x) represents the difference between the rates at which charge flows through the two x-oriented windows of the cell. The other two terms correspond to the rates of flow through the y and z-oriented windows. The sum of these three terms is the overall rate at which charge flows through all the boundaries of the box. In relativistic notation, the continuity equation (Eq. 10.24) becomes

$$\partial_\mu J^\mu = 0. \tag{10.25}$$

But how does this help us derive Maxwell's equations? Here's how: Given the continuity equation, we can construct a gauge invariant scalar that involves both J^μ and A_μ. The new scalar is $J^\mu A_\mu$, which may be the simplest possible way to combine these two quantities. It doesn't *look* gauge invariant, but we'll see that it is. We consider each of these quantities to be a function of position. We'll use the single variable x to stand for all three space components.

Now let's think about the impact this new scalar would

have if we add it to the Lagrangian. The action would contain the additional term

$$Action_J = - \int d^4x J^\mu(x) A_\mu(x).$$

The minus sign is a convention, ultimately due to Benjamin Franklin. $J^\mu(x)A_\mu(x)$ is a scalar because it's the contraction of a 4-vector with another 4-vector. It involves both the current density and the vector potential. How can we tell if it's gauge invariant? Simple: Just do a gauge transformation and see what happens to the action. Doing a gauge transformation means adding the gradient of some scalar to A_μ. The gauge-transformed action integral is

$$Action_J = - \int d^4x J^\mu(x) \left(A_\mu(x) + \frac{\partial S}{\partial X^\mu} \right).$$

What we really care about here is the *change* in action due to the extra term. That change is

$$Change\ in\ Action = - \int d^4x J^\mu(x) \frac{\partial S}{\partial X^\mu}.$$

This doesn't look much like zero. If it's not zero, then the action is not gauge invariant. But it is zero. Let's see why.

It helps to remember that d^4x is shorthand for $dt\ dx\ dy\ dz$. Let's expand the summation over μ. I'll stop writing the left side at this point because we don't need it. The expanded integral is

$$- \int d^4x \left(J^0 \frac{\partial S}{\partial X^0} + J^1 \frac{\partial S}{\partial X^1} + J^2 \frac{\partial S}{\partial X^2} + J^3 \frac{\partial S}{\partial X^3} \right). \quad (10.26)$$

We're going to make one key assumption: that if you go far enough away, there is no current. All currents in the problem are contained within a big laboratory so that every component of J goes to zero at large distances. If we come across a situation where there is a nonzero current at infinity, we have to treat it as a special case. But for any ordinary experiment, we can imagine that the laboratory is isolated and sealed, and that no currents exist outside it.

Let's look at one particular term in the expanded integral, the term

$$- \int J^1 \frac{\partial S}{\partial X^1} \, d^4x,$$

which we'll now write as

$$- \int J^1 \frac{\partial S}{\partial x} \, d^4x.$$

If you've read the previous books in this series, you already know where this is going. We're about to use an important technique called integration by parts. Because we're considering only the x component for now, we can treat this term as an integral over x and ignore the $dy \, dz \, dt$ portion of d^4x. To integrate by parts, we switch the derivative to the other factor and change the overall sign.[6] In other words, we can rewrite the integral as

$$\int \frac{\partial J^1}{\partial x} S \, d^4x.$$

What happens if we do the same thing with the next term in Eq. 10.26? The next term is

$$J^2 \frac{\partial S}{\partial X^2}.$$

---

[6] In general, integration by parts involves an additional term called the *boundary term*. The assumption that J goes to zero at great distances allows us to ignore the boundary term.

Because this term has the same mathematical form as the J^1 term, we get a similar result. In fact, all four terms in Eq. 10.26 follow this pattern, and we can capture that idea nicely using the summation convention. We can reformulate Eq. 10.26 as

$$\int \frac{\partial J^\mu}{\partial X^\mu} \, S \, d^4x$$

or

$$\int \partial_\mu J^\mu \, S \, d^4x.$$

Does the sum $\partial_\mu J^\mu$ in this integral look familiar? It should. It's just the left side of the continuity equation (Eq. 10.25). If the continuity equation is correct then this term, and in fact the entire integral, must be zero. If the current satisfies the continuity equation—and *only* if it satisfies the continuity equation—then adding the peculiar-looking term

$$Change\ to\ Lagrangian = J^\mu(x)A_\mu(x) \tag{10.27}$$

to the Lagrangian is gauge invariant.

How does this new term affect the equations of motion? Let's quickly review our derivation for empty space. We started with the Euler-Lagrange equation,

$$\frac{\partial}{\partial X^\mu} \frac{\partial \mathcal{L}}{\partial A_{\nu,\mu}} = \frac{\partial \mathcal{L}}{\partial A_\nu}.$$

We then filled in the details of this equation for each component of the field A. For the left side the result was

$$\frac{\partial F^{\mu\nu}}{\partial X^\nu}.$$

The right side, of course, was zero. This result is based on the original Lagrangian for empty space, Eq. 10.17.

The new term (Eq. 10.27) involves A itself but does not involve any derivatives of A. Therefore, it has no impact on the left side of the Euler-Lagrange equation. What about the right side? When we differentiate Eq. 10.27 with respect to A_μ, we just get J^μ. The equation of motion becomes

$$\frac{\partial F^{\mu\nu}}{\partial X^\nu} = J^\mu. \qquad (10.28)$$

These are the Maxwell equations we've been looking for. They are, of course, the four equations that constitute the second set of Maxwell equations,

$$\vec{\nabla} \cdot \vec{E} = \rho$$

$$c^2 \vec{\nabla} \times \vec{B} - \frac{\partial}{\partial \vec{E}} = j.$$

To summarize: We've found that Maxwell's equations do follow from an action or Lagrangian formulation. What is more, the Lagrangian is gauge invariant, but only if the current 4-vector satisfies the continuity equation. What would happen if the current failed to satisfy continuity? The answer is that the equations would simply not be consistent. We can see the inconsistency from Eq. 10.28 by differentiating both sides,

$$\frac{\partial^2 F^{\mu\nu}}{\partial X^\mu \partial X^\nu} = \frac{\partial J^\mu}{\partial X^\mu}.$$

The right side is just the expression that appears in the continuity equation. The left side is automatically zero because $F^{\mu\nu}$ is antisymmetric. If the continuity equation is not satisfied, the equations are contradictory.

Lecture 11

Fields and Classical Mechanics

"I'm annoyed, Lenny. You always go on and on about how our equations have to come from an action principle. Okay, you convinced me that action is an elegant idea, but as someone said, 'Elegance is for tailors.' Frankly, I don't see why we need it."

"Don't be grumpy, Art. There really is a reason. I bet that if I let you make up your own equations, energy won't be conserved. A world without energy conservation would be pretty weird. The sun could suddenly go out, or cars could spontaneously start moving for no reason."

"Okay, I get it: Lagrangians lead to conservation laws. I remember that from classical mechanics. Noether's theorem, right? But we've hardly mentioned energy and momentum conservation. Do electromagnetic fields really have momentum?"

"Yup. And Emmy can prove it."

11.1 Field Energy and Momentum

Electromagnetic fields clearly have energy. Stand in the sun for a few minutes; the warmth that you feel is the energy of sunlight being absorbed by your skin. It excites molecular motion and becomes heat. Energy is a conserved quantity, and it's carried by electromagnetic waves.

Electromagnetic fields also carry momentum, although it's not as easy to detect. When sunlight is absorbed by your skin, some momentum is transferred and it exerts a force or pressure on you. Fortunately the force is feeble, and standing in the sunlight won't exert much of a push. But it's there. Future space travel may utilize the pressure of sunlight (or even starlight) on a large sail in order to accelerate a spaceship (Fig. 11.1). Whether or not this is practical, the effect is real. Light carries momentum, which, when it's absorbed, exerts a force.[1]

Energy and momentum are the key concepts that connect field theory to classical mechanics. We'll start by having a closer look at what we actually mean by the words *energy* and *momentum*.

In this lecture I am going to assume you are familiar with the concepts of classical mechanics as they were presented in *Volume I* of the Theoretical Minimum.

11.2 Three Kinds of Momentum

We have encountered three different concepts of momentum. These are not three ways of thinking about the same thing,

---

[1] See https://en.m.wikipedia.org/wiki/IKAROS for a description of IKAROS, an experimental spacecraft based on this concept.

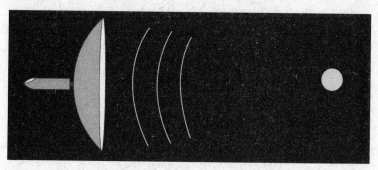

Figure 11.1: Solar Sail. Light waves carry momentum. They exert pressure on material objects and cause them to accelerate.

but three ways to think about three different things—things that in general have different numerical values. The first and simplest concept is mechanical momentum.

11.2.1 Mechanical Momentum

In non-relativistic physics mechanical momentum, called $\vec{p}$, is just mass times velocity. More precisely, it is the total mass of a system times the velocity of the center of mass. It's a vector quantity with three space components, p_x, p_y, and p_z. For a single particle, the components are

$$
\begin{aligned}
p_x &= m\dot{x} \\
p_y &= m\dot{y} \\
p_z &= m\dot{z}.
\end{aligned}
\tag{11.1}
$$

For a collection of particles labeled i the components of mechanical momentum are

$$
p_x = \sum_i m_i \dot{x}_i,
\tag{11.2}
$$

and similarly for the y and z components. Relativistic particles also have mechanical momentum given by

$$p_x = \frac{m\dot{x}}{\sqrt{1 - v^2/c^2}}. \tag{11.3}$$

11.2.2 Canonical Momentum

Canonical momentum is an abstract quantity that can apply to any kind of degree of freedom. For any coordinate that appears in the Lagrangian,[2] there is a canonical momentum. Suppose the Lagrangian depends on a set of abstract coordinates q_i. These coordinates could be the spatial coordinates of a particle, or the angle of a rotating wheel. They could even be the fields representing the degrees of freedom of a field theory. Each coordinate has a *conjugate canonical momentum* that is often denoted by the symbol Π_i. Π_i is defined as the derivative of the Lagrangian with respect to $\dot{q}_i$:

$$\Pi_i = \frac{\partial L(q_i, \dot{q}_i)}{\partial \dot{q}_i}.$$

If the coordinate q_i happens to be called x, and the Lagrangian happens to be

$$L = \frac{1}{2}m\dot{x}^2 - V(x),$$

where $V(x)$ is a potential energy function, then the canonical momentum is the same as the mechanical momentum.

Even if the coordinate in question represents the position of a particle, the canonical momentum may not be the mechanical momentum. In fact we've already seen an

---

[2] For now I'll use the symbol L instead of $\mathcal{L}$ for the Lagrangian.

example in Lecture 6. Sections 6.3.2 and 6.3.3 describe the motion of a charged particle in an electromagnetic field. The Lagrangian (from Eq. 6.18) is

$$L = -m\sqrt{1 - \dot{x}^2} + eA_0(x) + e\dot{X}^p A_p(x),$$

and you can see that velocity appears in more than one term. The canonical momentum (from Eq. 6.20) is

$$\frac{\partial L}{\partial \dot{X}_p} = m\frac{\dot{X}_p}{\sqrt{1 - \dot{x}^2}} + eA_p(x).$$

We can call this quantity Π_p. The first term on the right side is the same as the (relativistic) mechanical momentum. But the second term has nothing to do with mechanical momentum. It involves the vector potential and it's something new. When we develop classical mechanics in Hamiltonian form, we always use canonical momentum.

In many cases the coordinates describing a system have nothing to do with the positions of particles. A field theory is described by a set of fields at each point of space. For example, one simple field theory has Lagrangian density

$$L = \frac{1}{2}\left\{(\partial_t\phi)^2 - (\partial_x\phi)^2\right\}.$$

The canonical momentum conjugate to $\phi(x)$ in this theory is

$$\Pi(x) = \frac{\partial L}{\partial \dot{\phi}} = \dot{\phi}. \tag{11.4}$$

This "field momentum" is only distantly related to the usual concept of mechanical momentum.

11.2.3   Noether Momentum[3]

Noether momentum is related to symmetries. Let us suppose that a system is described by a set of coordinates or degrees of freedom q_i. Now suppose we change the configuration of the system by shifting the coordinates a tiny bit. We might write this in the form

$$q_i \rightarrow q_i + \delta_i. \tag{11.5}$$

The small shifts δ_i may depend on the coordinates. A good general way to write this is

$$\delta_i = \epsilon f_i(q), \tag{11.6}$$

where ϵ is an infinitesimal constant and the $f_i(q)$ are functions of the coordinates.

The simplest example is a translation of a system in space. A system of particles (labeled n) may be uniformly displaced along the x axis by amount ϵ. We express this by the equations

$$\delta X_n = \epsilon$$

$$\delta Y_n = 0$$

$$\delta Z_n = 0.$$

Each particle is shifted along the x axis by amount ϵ, while remaining at the same values of y and z. If the potential energy of the system only depends on the distances between particles, then the value of the Lagrangian does not change when the system is displaced in this way. In that case we say that there is translation symmetry.

---

[3] I use the term *Noether Momentum* because I'm not aware of a standard term for this quantity.

Another example would be a rotation around the origin in two dimensions. A particle with coordinates X, Y would be shifted to $X + \delta X$, $Y + \delta Y$, where

$$\delta X = -\epsilon Y$$
$$\delta Y = \epsilon X. \tag{11.7}$$

You can check that this corresponds to a rotation of the particle around the origin by angle ϵ. If the Lagrangian does not change, then we say that the system has rotation invariance.

A transformation of coordinates that does not change the value of the Lagrangian is called a symmetry operation.[4] According to Noether's theorem, if the Lagrangian doesn't change under a symmetry operation, then there is a conserved quantity that I'll call Q.[5] This quantity is the third concept of momentum. It may or may not be equal to mechanical or canonical momentum. Let's remind ourselves what it is.

We can express a coordinate shift with the equation

$$q'_i = q_i + \delta q_i = q_i + \epsilon f_i(q), \tag{11.8}$$

where δq_i represents an infinitesimal change to the coordinate q_i, and the function $f_i(q)$ depends on *all* of the q's, not just q_i. If Eq. 11.8 is a symmetry, then Noether's

---

[4] In this context, we're talking about an *active* transformation, which means we move the whole laboratory, including all the fields and all the charges, to a different location in space. This is different from a passive transformation that just relabels the coordinates.

[5] I explain Noether's theorem in the first volume of this series, *Classical Mechanics*. For more about Noether's contributions, see https://en.m.wikipedia.org/wiki/Emmy_Noether.

theorem tells us that the quantity

$$Q = \sum_i \Pi_i f_i(q) \qquad (11.9)$$

is conserved. Q is a sum over all the coordinates; there's contribution from each of the canonical momenta Π_i.

Let's consider a simple example where q happens to be the x position of a single particle. When you translate coordinates, x changes by a small amount that is independent of x. In that case δq (or δx) is just a constant; it's the amount by which you shift. The corresponding f is trivially just 1. The conserved quantity Q contains one term, which we can write as

$$Q = \Pi f(q). \qquad (11.10)$$

Since $f(q) = 1$, we see that Q is just the canonical momentum of the system. For the case of a simple nonrelativistic particle it is just the ordinary momentum.

11.3   Energy

Momentum and energy are close relatives; indeed they are the space and time components of a 4-vector. It should not be surprising that the law of momentum conservation in relativity theory means the conservation of all four components.

Let's recall the concept of energy in classical mechanics. The energy concept becomes important when the Lagrangian is invariant under a translation of the time coordinate t. Shifting the time coordinate plays the same role for energy as shifting the space coordinate does for momentum. Shifting the time coordinate means that "t" becomes "t plus a constant." Invariance of the Lagrangian under a time translation means that the answer to a question about an experiment doesn't depend on when the experiment starts.

Given a Lagrangian $L(q_i, \dot{q}_i)$ that's a function of q_i and $\dot{q}_i$, there's a quantity called the Hamiltonian, defined as

$$H = \sum_i p_i \dot{q}_i - L. \qquad (11.11)$$

The Hamiltonian is the energy, and for an isolated system it's conserved. Let's return to our simple example where q is the x position of a single particle, and the Lagrangian is

$$L = \frac{1}{2} m \dot{x}^2 - V(x). \qquad (11.12)$$

What is the Hamiltonian for this system? The canonical momentum is the derivative of the Lagrangian with respect to the velocity. In this example, the velocity $\dot{x}$ appears only in the first term and the canonical momentum p_i is

$$p_i = m\dot{x}.$$

Multiplying p_i by $\dot{q}_i$ (which becomes $\dot{x}$ in this example) the result is $m\dot{x}^2$. Next, we subtract the Lagrangian, resulting in

$$H = m\dot{x}^2 - \left[\frac{1}{2} m \dot{x}^2 - V(x) \right]$$

or

$$H = \frac{1}{2} m \dot{x}^2 + V(x).$$

We recognize this as the sum of kinetic energy and potential energy. Is it always that easy?

The simple Lagrangian of Eq. 11.12 has one term that depends on $\dot{x}^2$, and another term that does not contain $\dot{x}$ at all. Whenever a Lagrangian is nicely separated in this way, it's easy to identify the kinetic energy with the $\dot{x}^2$ terms, and the potential energy with the other terms. When the Lagrangian takes this simple form—squares of velocities minus things that don't depend on velocities—you can read off the Hamiltonian quickly without doing any extra work; just flip the signs of the terms that don't contain velocities.

If Lagrange's equations apply to everything on the right side of Eq. 11.11, then the Hamiltonian is conserved; it does not change with time.[6] Whether or not the Lagrangian has such a simple form, the total energy of a system is defined to be its Hamiltonian.

11.4   Field Theory

Field theory is a special case of ordinary classical mechanics. Their close connection is slightly obscured when we try to write everything in relativistic fashion. It's best to start out by giving up on the idea of making all equations explicitly invariant under Lorentz transformation. Instead, we'll choose a specific reference frame with a specific time coordinate and work within that frame. Later on, we'll address the issue of switching from one frame to another.

In classical mechanics, we have a time axis and a collection of coordinates called $q_i(t)$. We also have a principle of stationary action, with action defined as the time integral of a Lagrangian,

$$Action = \int L(q_i, \dot{q}_i)\,dt.$$

The Lagrangian itself depends on all the coordinates and their time derivatives. That's all of it: a time axis, a set of time-dependent coordinates, an action integral, and the principle of stationary action.

---

[6] For a full explanation, see *Volume I* of the Theoretical Minimum series.

11.4.1 Lagrangian for Fields

In field theory we also have a time axis, and coordinates or degrees of freedom that depend on time. But what are those coordinates?

For a single field, we can view the field variable ϕ as a *set* of coordinates. That seems odd, but remember that the field variable ϕ depends not only on time but also on position. It's the *nature* of this dependency on position that sets it apart from the q_i variables that characterize a particle in classical mechanics.

Let's imagine, hypothetically, that the ϕ dependence on position is discrete rather than continuous. Fig. 11.2 illustrates this idea schematically. Each vertical line represents a single degree of freedom, such as ϕ_1, ϕ_2, and so forth. We refer to them collectively as ϕ_i. This naming convention mimics the coordinate labels (such as q_i) of classical mechanics. We do this to emphasize two ideas:

1. Each ϕ_i is a separate and independent degree of freedom.
2. The index i is just a label that identifies a particular degree of freedom.

In practice, the field variable ϕ is not labeled by a discrete index i but by a continuous variable x, and we use the notation $\phi(t, x)$. However, we'll continue to think of x as a *label* for an independent degree of freedom, and not as a system coordinate. The field variable $\phi(t, x)$ represents an independent degree of freedom for each value of x.

We can see this even more concretely with a physical model. Fig. 11.3 shows a linear array of masses connected by springs. The masses can only move in the horizontal direction, and the motion of each mass is a separate degree of freedom, q_i, labeled with discrete indices i. As we pack more

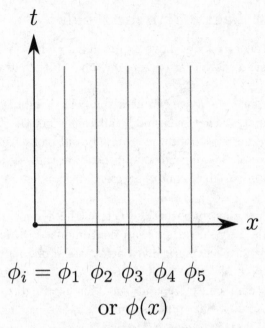

$$\phi_i = \phi_1 \ \ \phi_2 \ \ \phi_3 \ \ \phi_4 \ \ \phi_5$$
$$\text{or } \phi(x)$$

Figure 11.2: Elements of Field Theory. If we pretend that ϕ is discrete, we can think of ϕ_i in the same way we think of q_i in classical mechanics. The subscripts i are labels for independent degrees of freedom.

and more tiny masses and springs into the same space, this system starts to resemble a continuous mass distribution. In the limit, we would label the degrees of freedom by the continuous variable x rather than a discrete set of indices i. This scheme works for classical fields because they're continuous.

What about derivatives? Just as we expect, field Lagrangians depend on the time derivatives $\dot{\phi}$ of the field variables. But they also depend on derivatives of ϕ with respect to

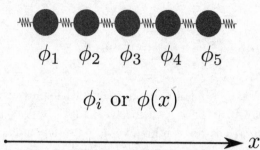

Figure 11.3: Field Analogy. Think of a set of discrete degrees of freedom such as these masses connected by springs. We label each degree of freedom with a subscript ϕ_i. Now suppose the masses become smaller and more densely spaced. In the limit there is an infinite number of small masses spaced infinitesimally closely. In this limit they become continuous (just like a field) rather than discrete, and it makes more sense to label the degrees of freedom $\phi(x)$ rather than ϕ_i.

space. In other words, they depend on quantities such as

$$\frac{\partial \phi}{\partial x}.$$

This is different from classical particle mechanics, where the only derivatives in the Lagrangian are time derivatives such as $\dot{q}_i$. A Lagrangian for field theory typically depends on things like

$$\phi(t, x),$$

$$\dot{\phi}(t, x),$$

and

$$\frac{\partial \phi(t, x)}{\partial x}.$$

But what is the derivative of ϕ with respect to x? It's defined as

$$\frac{\phi(x + \epsilon) - \phi(x)}{\epsilon}$$

for some small value of ϵ. In this sense, the space derivatives are functions of the ϕ's themselves; most importantly, the space derivatives do not involve $\dot{\phi}$. The dependence of the Lagrangian on space derivatives reflects its dependence on the ϕ's themselves. In this case, it depends on two nearby ϕ's at the same time.

11.4.2   Action for Fields

Let's think about what we mean by action. An action in classical mechanics is always an integral over time,

$$Action = \int dt L\left(\phi, \dot{\phi}, \frac{\partial \phi}{\partial x}\right).$$

But for fields, the Lagrangian itself is an integral over space. We've already seen this in previous examples. If we separate the time portion of the integral from the space portion, we can write the action as

$$
\begin{aligned}
Action &= \int dt L\left(\phi, \dot{\phi}, \frac{\partial \phi}{\partial x}\right) \\
&= \int dt \int d^3x \mathcal{L}\left(\phi, \dot{\phi}, \frac{\partial \phi}{\partial x}\right).
\end{aligned}
\tag{11.13}
$$

I'm using the symbol L for the Lagrangian, and the symbol $\mathcal{L}$ for the Lagrangian *density*. The integral of $\mathcal{L}$ over space is the same as L. The point of this notation is to avoid mixing up the time derivatives with the space derivatives.

11.4.3 Hamiltonian for Fields

The energy of a field is part of the universal conserved energy that is associated with time translation. To understand field energy, we need to construct the Hamiltonian. To do that, we need to identify the generalized coordinates (the q's), their corresponding velocities and canonical momenta (the $\dot{q}$'s and the p's), and the Lagrangian. We've already seen that the coordinates are $\phi(t, x)$. The corresponding velocities are just $\dot{\phi}(t, x)$. These are not velocities in space, but quantities that tell us how fast the field changes at a particular point in space. The canonical momentum conjugate to ϕ is the derivative of the Lagrangian with respect to $\dot{\phi}$. We can write this as

$$\Pi_\phi(x) = \frac{\partial \mathcal{L}}{\partial \dot{\phi}}(x),$$

where $\Pi_\phi(x)$ is the canonical momentum conjugate to ϕ. Both sides of the equation are functions of position. If there are many fields in a problem—such as ϕ_1, ϕ_2, ϕ_3—there will be a different $\Pi(x)$ associated with each of them. With these preliminaries in place, we're able to write the Hamiltonian. The Hamiltonian defined by Eq. 11.11,

$$H = \sum_i p_i \dot{q}_i - L,$$

is a sum over i. But what does i represent in this problem? It labels a degree of freedom. Because fields are continuous, their degrees of freedom are instead labeled by a real variable x, and the sum over i becomes an integral over x. If we replace p_i with $\Pi_\phi(x)$, and $\dot{q}_i$ with $\dot{\phi}(x)$, we have the correspondence

$$\sum_i p_i \dot{q}_i \Longrightarrow \int d^3 x \Pi_\phi(x) \dot{\phi}(x).$$

To turn this integral into a Hamiltonian, we have to subtract the total Lagrangian L. But we know from Eq. 11.13 that the Lagrangian is

$$L = \int d^3x \mathcal{L}\left(\phi, \dot{\phi}, \frac{\partial \phi}{\partial x}\right).$$

This is also an integral over space. Therefore, we can put both terms of the Hamiltonian inside the same integral, and the Hamiltonian becomes

$$H = \int d^3x \left[\Pi_\phi(x)\dot{\phi}(x) - \mathcal{L}\right].$$

This equation is interesting; it expresses the energy as an integral over space. Therefore, the integrand itself is an *energy density*. This is a characteristic feature of all field theories; conserved quantities like energy, and also momentum, are integrals of densities over space.

Let's go back to the Lagrangian of Eq. 4.7, which we've used on a number of occasions. Here, we view it as a Lagrangian density. We'll consider a simplified version, with only a single dimension of space,

$$\mathcal{L} = \frac{1}{2}\dot{\phi}^2 - \frac{1}{2}\left(\frac{\partial \phi}{\partial X}\right)^2 - V(\phi). \qquad (11.14)$$

The first term is the kinetic energy.[7] What is Π_ϕ? It's the derivative of $\mathcal{L}$ with respect to $\dot{\phi}$, which is just $\dot{\phi}$. That is,

$$\Pi_\phi = \dot{\phi},$$

---

[7] If we went back to thinking about relativity, we would recognize the first two terms on the right side as scalars, and combine them into a single term such as $\frac{1}{2}\partial_\mu\phi\partial^\mu\phi$. That's a beautiful relativistic expression, but we don't want to use this form just now. Instead, we'll keep track of time derivatives and space derivatives separately.

and the Hamiltonian, (the energy), is

$$H = \int dx \left[\Pi_\phi \dot\phi - \mathcal{L} \right].$$

Replacing Π_ϕ with $\dot\phi$, this becomes

$$H = \int dx \left[(\dot\phi)^2 - \mathcal{L} \right].$$

Plugging in the expression for $\mathcal{L}$ gives

$$H = \int dx \left[(\dot\phi)^2 - \left\{ \frac{1}{2}(\dot\phi)^2 - \frac{1}{2}\left(\frac{\partial\phi}{\partial x}\right)^2 - V(\phi) \right\} \right]$$

or

$$H = \int dx \left[\frac{1}{2}(\dot\phi)^2 + \frac{1}{2}\left(\frac{\partial\phi}{\partial x}\right)^2 + V(\phi) \right]. \qquad (11.15)$$

In Eq. 11.14 (the Lagrangian density), we have a kinetic energy term containing the time derivative $\dot\phi^2$. The second two terms, containing $V(\phi)$ and the space derivative of ϕ, play the role of potential energy—the part of the energy that does not contain time derivatives.[8] The Hamiltonian in this example consists of kinetic energy *plus* potential energy and represents the total energy of the system. By contrast, the Lagrangian is kinetic energy *minus* potential energy.

Let's set the $V(\phi)$ term aside for a moment and consider the terms involving time and space derivatives. Both of these terms are positive (or zero) because they're squares. The Lagrangian has both positive and negative terms, and is not

---

[8] Sometimes $V(\phi)$ is called the field potential energy, but it's more accurate to think of the combination of both terms as the potential energy. Potential energy should include any terms that don't contain time derivatives.

necessarily positive. But energy has only non-negative terms. Of course these terms can be zero. But the only way that can happen is for ϕ to be constant; if ϕ is constant then its derivatives must be zero. It should be no surprise that energy does not typically become negative.

What about $V(\phi)$? This term can be positive or negative. But if it's bounded from below, then we can easily arrange for it to be positive by adding a constant. Adding a constant to $V(\phi)$ doesn't change anything about the equations of motion. However, if $V(\phi)$ is not bounded from below, the theory is not stable and everything goes to hell in a handwagon; you don't want any part of such a theory. So we can assume that $V(\phi)$, and consequently the total energy, is zero or positive.

11.4.4   Consequences of Finite Energy

If x is just a label and $\phi(t, x)$ is an independent degree of freedom for each value of x, what prevents ϕ from varying wildly from one point to another? Is there any requirement for $\phi(t, x)$ to vary smoothly? Suppose hypothetically that the opposite is true: namely, that the value of ϕ may jump sharply between neighboring points in Fig. 11.2. In that case, the gradient (or space derivative) would get huge as the separation between points gets smaller.[9]

This means the energy density would become infinite as the separation decreased. If we're interested in configurations where the energy doesn't blow up to infinity, these derivatives must be finite. Finite derivatives tell $\phi(t, x)$ to be

---

[9] Remember what the derivative is. It's the change in ϕ between two nearby points, divided by the small separation. For a given change in ϕ, the smaller the separation, the bigger the derivative.

smooth; they put the brakes on how wildly ϕ can vary from point to point.

11.4.5 Electromagnetic Fields via Gauge Invariance

How can we apply these ideas about energy and momentum to the electromagnetic field? We could certainly use the Lagrangian formulation that we saw earlier. The fields are the four components of the vector potential A_μ. The field tensor $F^{\mu\nu}$ is written in terms of the space and time derivatives of these components. We then square the components of $F^{\mu\nu}$ and add them up to get the Lagrangian. Instead of having just one field, we would have four.[10]

But there is a helpful simplification based on gauge invariance. This simplification not only makes our work easier but illustrates some important ideas about gauge invariance. Remember that the vector potential is not unique; if we're clever, we can change it in ways that do not affect the physics. This is similar to the freedom to choose a coordinate system, and we can use that freedom to simplify our equations. In this case, our use of gauge invariance will allow us to work with only three components of the vector potential rather than worrying about all four.

Let's recall what a gauge transformation is; it's a transformation that adds the gradient of an *arbitrary* scalar S to the vector potential A_μ. It amounts to the replacement

$$A_\mu \Longrightarrow A_\mu + \frac{\partial S}{\partial X^\mu}.$$

---

[10] We'll work mainly in relativistic units ($c = 1$). On occasion, we'll briefly restore the speeds of light when a sense of relative scale is important.

We can use this freedom to pick S in a way that simplifies A_μ. In this case we'll choose S in a way that makes the time component A_0 become zero. All we care about is the time derivative of S, because A_0 is the time component of A_μ. In other words, we'd like to choose an S such that

$$A_0 + \frac{\partial S}{\partial t} = 0$$

or

$$\frac{\partial S}{\partial t} = -A_0.$$

Is that possible? The answer is yes. Remember that $\frac{\partial S}{\partial t}$ is a derivative with respect to time at a fixed position in space. If you go to that fixed position in space, you can always choose S so that its time derivative is some prespecified function, $-A_0$. If we do this, then the new vector potential,

$$(A')_\mu = A_\mu + \frac{\partial S}{\partial t},$$

will have zero for its time component. This is called *fixing the gauge*. Different choices of gauges have names—the Lorentz gauge, the radiation gauge, the Coulomb gauge. The gauge with $A_0 = 0$ goes under the elegant name "the $A_0 = 0$ gauge." We could have made other choices; none of them would affect the physics, but the $A_0 = 0$ gauge is especially convenient for our purposes. Based on our choice of gauge, we write

$$A_0 = 0.$$

This completely eliminates the time component from the vector potential. All we have left are the space components,

$$A_m(x).$$

With this gauge in place, what are the electric and magnetic fields? In terms of the vector potential, the electric field is defined (see Eq. 8.7) as

$$\vec{E} = -\frac{\partial \vec{A}}{\partial t} + \vec{\nabla} A_0.$$

But with A_0 set equal to zero, the second term disappears and we can write a simpler equation,

$$\vec{E} = -\frac{\partial \vec{A}}{\partial t}. \tag{11.16}$$

The electric field is just the time derivative of the vector potential. What about the magnetic field? It only depends on the space components of the vector potential, and these are not affected by our choice of gauge. Thus it's still true that

$$\vec{B} = \vec{\nabla} \times \vec{A}. \tag{11.17}$$

This simplifies things because we can now ignore the time component of A_μ altogether. The degrees of freedom in the $A_0 = 0$ gauge are just the space components of the vector potential.

Let's consider the form of the Lagrangian in the $A_0 = 0$ gauge. In terms of the field tensor, recall (Eq. 10.17) that the Lagrangian is

$$\mathcal{L} = -\frac{1}{4} F_{\mu\nu} F^{\mu\nu}. \tag{11.18}$$

But (as we saw in Eq. 10.18) that happens to equal $\frac{1}{2}$ times the square of the electric field minus the square of the magnetic field,

$$\mathcal{L} = \frac{1}{2}(E^2 - B^2). \tag{11.19}$$

The first term in Eq. 11.19 is

$$\frac{1}{2}E^2.$$

But from Eq. 11.16 we know that this is the same as

$$\frac{1}{2}\left(\frac{\partial \vec{A}}{\partial t}\right)^2.$$

With this substitution, Eq. 11.19 is starting to resemble Eq. 11.14, which was

$$\mathcal{L} = \frac{\dot{\phi}^2}{2} - \frac{(\partial_x \phi)^2}{2} - V(\phi).$$

Notice that $\frac{1}{2}\left(\frac{\partial \vec{A}}{\partial t}\right)^2$ is not just a single term, but three terms; it contains the squares of the time derivatives of A_x, A_y, and A_z. It is a sum of terms of exactly the same type as the $\frac{\dot{\phi}^2}{2}$ term of Eq. 11.14—one term for each component of the vector potential. The expanded form is

$$\frac{1}{2}\left[\left(\frac{\partial A_x}{\partial t}\right)^2 + \left(\frac{\partial A_y}{\partial t}\right)^2 + \left(\frac{\partial A_z}{\partial t}\right)^2\right].$$

The second term of Eq. 11.19 is the square of the curl of A. The full Lagrangian density is

$$\mathcal{L} = \frac{1}{2}\left(\frac{\partial \vec{A}}{\partial t}\right)^2 - \frac{1}{2}(\vec{\nabla} \times \vec{A})^2 \qquad (11.20)$$

or

$$\mathcal{L} = \frac{1}{2}\left[\left(\frac{\partial A_x}{\partial t}\right)^2 + \left(\frac{\partial A_y}{\partial t}\right)^2 + \left(\frac{\partial A_z}{\partial t}\right)^2\right] - \frac{1}{2}(\vec{\nabla} \times \vec{A})^2. \qquad (11.21)$$

The resemblance to Eq. 11.14 is even stronger now: squares of time derivatives minus squares of space derivatives.

What is the canonical momentum conjugate to a particular component of the vector potential? By definition, it's

$$\Pi_A = \frac{\partial \mathcal{L}}{\partial(\partial_t A)},$$

which is just

$$\Pi_A = \frac{\partial A}{\partial t}.$$

In terms of individual components,

$$\Pi_x = \frac{\partial A_x}{\partial t}$$

$$\Pi_y = \frac{\partial A_y}{\partial t}$$

$$\Pi_z = \frac{\partial A_z}{\partial t}.$$

But it's also true that the time derivative of A is minus the electric field. So we found something interesting. The canonical momenta happen to be minus the components of the electric field. In other words,

$$\Pi_x = \frac{\partial A_x}{\partial t} = -E_x$$

$$\Pi_y = \frac{\partial A_y}{\partial t} = -E_y$$

$$\Pi_z = \frac{\partial A_z}{\partial t} = -E_z.$$

Thus the physical meaning of canonical momentum conjugate to the vector potential is (minus) the electric field.

What's the Hamiltonian?

Now that we know the Lagrangian, we can write down the Hamiltonian. We could go through the formal construction of the Hamiltonian, but in this case we don't need to. That's because our Lagrangian has the form of a kinetic energy term that depends on squares of time derivatives, minus a potential energy term that has no time derivatives at all. When the Lagrangian has this form—kinetic energy minus potential energy—we *know* what the answer is: The Hamiltonian is kinetic energy *plus* potential energy. Thus the electromagnetic field energy is

$$H = \frac{1}{2}(E^2 + B^2). \qquad (11.22)$$

Once again, the Lagrangian is not necessarily positive. In particular, if there's a magnetic field with no electric field, the Lagrangian is negative. But the energy, $\frac{1}{2}(E^2 + B^2)$, is positive. What does this say about an electromagnetic plane wave moving along an axis? The example we saw in Lecture 10 had an E component in one direction and a B component in the perpendicular direction. The B field has the same magnitude as the E field and is in phase with it, but polarized in the perpendicular direction. This tells us that the electric and magnetic field energies are the same. An electromagnetic wave moving down the z axis has both electric and magnetic energy, and both contributions happen to be the same.

Momentum Density

How much momentum does an electromagnetic wave carry? Let's go back to Noether's concept of momentum from Section 11.2.3. The first step in using Noether's theorem is to identify a symmetry. The symmetry associated with momentum conservation is translation symmetry along a spatial direction.

For example, we may translate a system along the x axis by a small distance ϵ (see Fig. 11.4). Each field $\phi(x)$ gets replaced by $\phi(x - \epsilon)$. Accordingly, the change in $\phi(x)$ is

$$\delta\phi = \phi(x - \epsilon) - \phi(x),$$

which (if ϵ is infinitesimal) becomes

$$\delta\phi = -\epsilon\frac{\partial\phi}{\partial x}.$$

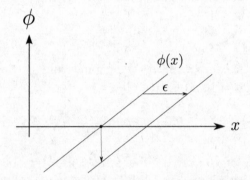

Figure 11.4: Noether Shift. Take a field conguration $\phi(x)$ and shift it to the right (toward higher x) by an infinitesimal amount ϵ. The change in ϕ at a specific point is $-\epsilon d\phi$.

In the case of electromagnetism the fields are the space components of the vector potential. The shifts of these fields under a translation become

$$\delta A_x = -\epsilon\frac{\partial A_x}{\partial x}$$

$$\delta A_y = -\epsilon\frac{\partial A_y}{\partial x}$$

$$\delta A_z = -\epsilon\frac{\partial A_z}{\partial x}. \tag{11.23}$$

Next we recall from Eq. 11.9 that the conserved quantity associated with this symmetry has the form

$$Q = \sum_i \Pi_i f_i(q).$$

In this case the canonical momenta are minus the electric fields

$$\Pi \to -E,$$

and the f_i are just the expressions multiplying ϵ in Eqs. 11.23,

$$f_i \to -\frac{\partial A_i}{\partial x}.$$

Thus the x component of the momentum carried by the electromagnetic field is given by

$$P_x = \int dx E_m \frac{\partial A_m}{\partial x}.$$

We could, of course, do the same thing for the y and z directions in order to get all three components of momentum. The result is

$$P_n = \int dx E_m \frac{\partial A_m}{\partial X^n}. \tag{11.24}$$

Evidently, like the energy, the momentum of an electromagnetic field is an integral over space.[11] We might therefore identify the integrand of Eq. 11.24 as the density of momen-

---

[11] Because these integrals reference one space component at a time, we've written dx instead of d^3x. A more precise notation for the integrand of Eq. 11.24 might be $dX^n E_m \frac{\partial A_m}{\partial X^n}$. We've decided on the simpler form, where dx is understood to reference the appropriate space component.

tum,

$$(\text{Momentum Density})_n = E_m \frac{\partial A_m}{\partial X^n}.$$

There is an interesting contrast between this momentum density and the energy density in Eq. 11.22. The energy density is expressed directly in terms of the electric and magnetic fields, but the momentum density still involves the vector potential. This is disturbing; the electric and magnetic fields are gauge invariant, but the vector potential is not. One might think that quantities like energy and momentum densities should be similar and only depend on E and B. In fact there is a simple fix that turns the momentum density into a gauge invariant quantity. The quantity

$$\frac{\partial A_m}{\partial X^n}$$

is part of the expression for the magnetic field. In fact, we can turn it into a magnetic field component by adding the term

$$-\frac{\partial A_n}{\partial X^m}.$$

If we could sneak in this change, we'd be able to rewrite the integral as

$$P_n = \int dx E_m \left(\frac{\partial A_m}{\partial X^n} - \frac{\partial A_n}{\partial X^m} \right).$$

Can we get away with it? How would P_n change if we insert this additional term? To find out, let's look at the term we want to add,

$$-\int dx E_m \frac{\partial A_n}{\partial X^m}.$$

Let's integrate this by parts. As we've seen before, whenever we take the integral whose integrand is a derivative times something else, we can shift the derivative to the other term at the cost of a minus sign:[12]

$$-\int dx E_m \frac{\partial A_n}{\partial X^m} = \int dx \frac{\partial E_m}{\partial X^m} A_n.$$

But due to the summation index m, the term $\frac{\partial E_m}{\partial X^m}$ is just the divergence of the electric field. We know from Maxwell's equations for empty space that $\vec{\nabla} \cdot \vec{E}$ is zero. That means the additional integral changes nothing, and therefore that P_n can be written in terms of the electric *and* magnetic fields,

$$P_n = \int dx E_m \left(\frac{\partial A_m}{\partial X^n} - \frac{\partial A_n}{\partial X^m} \right). \qquad (11.25)$$

A little algebra will show that the integrand is actually the vector $\vec{E} \times \vec{B}$. That means $\vec{E} \times \vec{B}$ is the momentum density. Dropping the subscripts and reverting to standard vector notation, Eq. 11.25 now becomes

$$\vec{P} = \int \vec{E} \times \vec{B} \; d^3x.$$

$\vec{P}$ is a vector, and its direction tells you the direction of the momentum. The momentum density $\vec{E} \times \vec{B}$ is called the *Poynting vector*, often denoted by $\vec{S}$.

The Poynting vector tells us something about how the wave propagates. Let's go back to Fig. 10.1 and look at the first two half-cycles of the propagating wave. In the first half-cycle, $\vec{E}$ points upward, and $\vec{B}$ points out of the page. Following the right-hand rule, we can see that $\vec{E} \times \vec{B}$ points to the right, along the z axis. That's the direction in which momentum (and the wave itself) propagates. What about the

---

[12] We assume that the fields go to zero beyond a certain distance, and therefore there are no boundary terms.

next half-cycle? Since the direction of both field vectors has been reversed, the Poynting vector still "poynts" to the right. We'll have more to say about this vector as we go.

Notice how important it is that $\vec{E}$ and $\vec{B}$ are perpendicular to each other; the momentum (the Poynting vector) is a consequence. This all traces back to Emmy Noether's wonderful theorem about the connection between conserved momentum and spatial translation invariance. It's all coming home in one big piece—classical mechanics, field theory, and electromagnetism. It all comes back to the principle of least action.

11.5 Energy and Momentum in Four Dimensions

We know that energy and momentum are conserved quantities. In four-dimensional spacetime, we can express this idea as a principle of local conservation. The ideas we present here are inspired by our previous work with charge and current densities and will draw on those results.

11.5.1 Locally Conserved Quantities

Back in Section 8.2.6, we explored the concept of local conservation of charge; if the charge within a region increases or decreases, it must do so by moving across the boundary of the region. Charge conservation is local in any sensible relativistic theory and gives rise to the concepts of charge density and current density, ρ and $\vec{j}$. Together, these two quantities make a 4-vector,

$$\rho, \vec{j} \implies J^{\mu}.$$

Charge density is the time component, and current density (or flux) is the space component.

The same concept applies to other conserved quantities. If we think more generally, beyond charges, we can imagine for each conservation law four quantities representing the density and flux of *any* conserved quantity. In particular, we can apply this idea to the conserved quantity called energy.

In Lecture 3, we learned that the energy of a particle is the time component of a 4-vector. The space components of that 4-vector are its relativistic momenta. Symbolically, we can write

$$E, \vec{p} \implies P^{\mu}.$$

In field theory, these quantities become densities, and energy density can be viewed as the time component of a four-dimensional current. We've already derived an expression for energy density; Eq. 11.22 gives the energy density of an electromagnetic field. Let's give this quantity the temporary name T^0:

$$T^0 = \frac{1}{2}\left(E^2 + B^2\right). \qquad (11.26)$$

What we want to do is find a current of energy analogous to the electric current J^m. Like J^m, the energy current has three components. Let's call them T^m. If we define T^m correctly, they should satisfy the continuity equation

$$\frac{\partial T^0}{\partial t} + \vec{\nabla} \cdot \vec{T} = 0.$$

Our strategy for finding the current of energy T^m is simple: Differentiate T^0 with respect to time, and see if the result is the divergence of something. Taking the time derivative of

Eq. 11.26 results in

$$\partial_t T^0 = \partial_t \left[\frac{1}{2} \left(E^2 + B^2 \right) \right]$$

or

$$\partial_t T^0 = \vec{E} \cdot \dot{\vec{E}} + \vec{B} \cdot \dot{\vec{B}}, \tag{11.27}$$

where $\dot{\vec{E}}$ and $\dot{\vec{B}}$ are derivatives with respect to time. At first glance, the right side of Eq. 11.27 doesn't look like a divergence of anything, because it involves time derivatives rather than space derivatives. The trick is to use the Maxwell equations

$$\dot{\vec{E}} = \vec{\nabla} \times \vec{B}$$

$$\dot{\vec{B}} = -\vec{\nabla} \times \vec{E} \tag{11.28}$$

to replace $\dot{\vec{E}}$ and $\dot{\vec{B}}$. Plugging these into Eq. 11.27 gives something a little more promising:

$$\partial_t T^0 = -\left[\vec{B} \cdot \left(\vec{\nabla} \times \vec{E} \right) - \vec{E} \cdot \left(\vec{\nabla} \times \vec{B} \right) \right]. \tag{11.29}$$

In this form, the right side contains space derivatives and therefore has some chance of being a divergence. In fact, using the vector identity

$$\vec{\nabla} \cdot \left(\vec{E} \times \vec{B} \right) = \vec{B} \cdot \left(\vec{\nabla} \times \vec{E} \right) - E \cdot \left(\vec{\nabla} \times \vec{B} \right),$$

we find that Eq. 11.29 becomes

$$\partial_t T^0 = -\vec{\nabla} \cdot \left(\vec{E} \times \vec{B} \right). \tag{11.30}$$

Lo and behold, if we define the current of energy as

$$\vec{T} = \vec{E} \times \vec{B}, \tag{11.31}$$

we can write the continuity equation for energy as

$$\frac{\partial T^0}{\partial t} + \vec{\nabla} \cdot \vec{T} = 0. \tag{11.32}$$

In relativistic notation, this becomes

$$\partial_\mu T^\mu = 0. \tag{11.33}$$

The vector $\vec{E} \times \vec{B}$ representing the flow of energy should be familiar; it's the Poynting vector, named after John Henry Poynting, who discovered it in 1884, probably by the same argument. We already met this vector in Section 11.4.5 under the heading "Momentum Density." The Poynting vector has two meanings: We can think of it as either an energy flow or a momentum density.

To summarize: Energy conservation is local. Just like charge, a change of energy within some region of space is always accompanied by a flow of energy through the boundaries of the region. Energy cannot suddenly disappear from our laboratory and reappear on the moon.

11.5.2   Energy, Momentum, and Lorentz Symmetry

Energy and momentum are not Lorentz invariant. That should be easy to see. Imagine an object of mass m at rest in your own rest frame. The object has energy given by the well-known formula

$$E = mc^2,$$

or in relativistic units,

$$E = m.$$

Because the object is at rest, it has no momentum.[13] But now look at the same object from another frame of reference where it is moving along the x axis. Its energy is increased, and now it does have momentum.

If field energy and momentum are not invariant, how do they change under a Lorentz transformation? The answer is that they form a 4-vector, just as they do for particles. Let's call the components of the 4-vector P^{μ}. The time component is the energy, and the three space components are ordinary momenta along the x, y, and z axes. All four components are conserved:

$$\frac{dP^{\mu}}{dt} = 0. \tag{11.34}$$

This suggests that each component has a density and that the total value of each component is an integral of the density. In the case of energy we denoted the density by the symbol T^0. But now we're going to change the notation by adding a second index and calling the energy density T^{00}.

I'm sure you've already guessed that the double index means we're building a new tensor. Each index has a specific meaning. The first index tells us which of the four quantities the element refers to.[14] The energy is the time component of the 4-momentum, and therefore the first index is 0 for time. To be explicit: A value of 0 for the first index indicates that we're talking about energy. A value of 1 indicates the x component of momentum. Values of 2 and 3 indicate the y and z components of momentum respectively.

The second index tells us whether we're talking about a

---

[13] In these two equations E stands for energy, not electric field.

[14] The roles of the first and second indices here may be reversed in relation to their roles described in the video. Because the tensor is symmetric, this makes no real difference.

density or a flow. A value of 0 indicates a density. A value of 1 indicates a flow in the x direction. Values of 2 and 3 indicate flows in the y and z directions respectively.

For example, T^{00} is the density of energy; we get the total energy by integrating T^{00} over space:

$$P^0 = \int T^{00} d^3x. \tag{11.35}$$

Now let's consider the x component of momentum. In this case the first index is x (or the number 1), indicating that the conserved quantity is the x component of momentum. The second index again differentiates between density and flow or current. Thus, for example,

$$P^1 = \int T^{10} d^3x$$

or more generally,

$$P^m = \int T^{m0} d^3x. \tag{11.36}$$

What about the *flow* of momentum? Each component has its own flux. For example, we can consider the flux of x-momentum flowing in the y direction.[15] This would be denoted T^{xy}. Similarly T^{zx} is the flux of z-momentum flowing in the x direction.

The trick in understanding $T^{\mu\nu}$ is to blot out the second index for a moment. The first index tells us which quantity

---

[15] The term *flux* may be confusing to some. Perhaps it's more clear if we call it the change of (the x component of momentum) in the y direction. Change in momentum should be a familiar idea; in fact, it represents force. But since we're talking about momentum *density*, it's better to think of these space-space components of $T^{\mu\nu}$ as stresses. The negatives of these space-space components form a 3×3 tensor in their own right, known as the *stress tensor*.

we are talking about, namely P^0, P^x, P^y, or P^z. Then, once we know which quantity, we blot out the first index and look at the second. That tells us if we are talking about a density or a component of the current.

We can now write down the continuity equation for the component of momentum P^m as

$$\frac{\partial T^{m0}}{\partial t} + \frac{\partial T^{mn}}{\partial X^n} = 0$$

or

$$\frac{\partial T^{m\nu}}{\partial X^\nu} = 0. \tag{11.37}$$

There are three such equations, one for each component of momentum (that is, one for each value of m). But if we add to these three the fourth equation representing the conservation of energy (by replacing m with μ) we can write all four equations in a unified relativistic form,

$$\frac{\partial T^{\mu\nu}}{\partial X^\nu} = 0. \tag{11.38}$$

This all takes some getting used to, so it might be a good time to stop and review the arguments. When you're ready to continue, we'll work out the expressions for momenta and their fluxes in terms of the fields E and B.

11.5.3   The Energy-Momentum Tensor

There are lots of ways to figure out what $T^{\mu\nu}$ is in terms of the electric and magnetic fields. Some are more intuitive than others. We're going to use a type of argument that may seem somewhat less intuitive and more formal, but it's common in modern theoretical physics and is also very powerful. What I have in mind is a symmetry or invariance argument. An invariance argument begins with listing the various symmetries of a system and then asking how the quantity of interest transforms under those symmetries.

The most important symmetries of electrodynamics are gauge invariance and Lorentz invariance. Let's begin with gauge invariance: How do the components of $T^{\mu\nu}$ transform under a gauge transformation? The answer is simple; they don't. The densities and fluxes of energy and momentum are physical quantities that must not depend on the choice of gauge. This means that they should depend only on the gauge-invariant observable fields $\vec{E}$ and $\vec{B}$ and not have additional dependence on the potential A_μ.

Lorentz invariance is more interesting. What kind of object is $T^{\mu\nu}$, and how do the components change in going from one frame of reference to another? It is clearly not a scalar because it has components labeled μ and ν. It cannot be a 4-vector because it has two indices and sixteen components. The answer is obvious. $T^{\mu\nu}$ is a tensor—a rank two tensor, meaning that it has two indices.

Now we can give $T^{\mu\nu}$ its proper name: the *energy-momentum tensor*.[16] Because it is such an important object, not just in electrodynamics but in all field theories, I'll repeat it: $T^{\mu\nu}$ is the *energy-momentum* tensor. Its components are the densities and currents of energy and momentum.

We can build $T^{\mu\nu}$ by combining the components of the field tensor $F^{\mu\nu}$. In general, there are many tensors we could build in this way, but we already know exactly what T^{00} is. It's the energy density,

$$T^{00} = \frac{1}{2}\left(E^2 + B^2\right). \qquad (11.39)$$

This tells us that $T^{\mu\nu}$ is quadratic in the components of the

---

[16] Some authors call it the *stress-energy* tensor.

field tensor; in other words, it is formed from products of two components of $F^{\mu\nu}$.

The question then is, how many different ways are there to form a tensor from the product of two copies of $F^{\mu\nu}$? Fortunately there are not too many. In fact, there are only two. Any tensor built quadratically out of $F^{\mu\nu}$ must be a sum of two terms of the form

$$T^{\mu\nu} = a \, F^{\mu\sigma} F^\nu_{\ \sigma} + b \, \eta^{\mu\nu} F^{\sigma\tau} F_{\sigma\tau}, \qquad (11.40)$$

where a and b are numerical constants that we'll figure out in a moment.

Let's pause to look at Eq. 11.40. The first thing to notice is that we've made use of the Einstein summation convention. In the first term the index σ is summed over, and in the second term σ and τ are summed over.

The second thing to notice is the appearance of the metric tensor $\eta^{\mu\nu}$. The metric tensor is diagonal and has components $\eta^{00} = -1$, $\eta^{11} = \eta^{22} = \eta^{33} = +1$. The only question is how to determine the numerical constants a and b.

Here the trick is to realize that we already know one of the components: T^{00} is the energy density. All we need to do is use Eq. 11.40 to determine T^{00} and then plug it into Eq. 11.39. Here is what we get:

$$aE^2 - b(2B^2 - 2E^2) = \frac{1}{2}\left(E^2 + B^2\right). \qquad (11.41)$$

Comparing the two sides, we find that $a = 1$ and $b = -1/4$. With these values of a and b, Eq. 11.40 becomes

$$T^{\mu\nu} = F^{\mu\sigma} F^\nu_{\ \sigma} - \frac{1}{4} \, \eta^{\mu\nu} F^{\sigma\tau} F_{\sigma\tau}. \qquad (11.42)$$

From this equation we can compute all the various components of the energy-momentum tensor.

Aside from T^{00}, which is given by Eq. 11.39, the most

interesting components are T^{0n} and T^{n0}, where n is a space index. T^{0n} are the components of the flux (or current) of energy, and if we work them out we find that they are (as expected) the components of the Poynting vector.

Now let's look at T^{n0}. Here we can use an interesting property of Eq. 11.40. A little bit of inspection should convince you that $T^{\mu\nu}$ is symmetric. In other words,

$$T^{\mu\nu} = T^{\nu\mu}.$$

Thus T^{n0} is the same as T^{0n}; both are just the Poynting vector. But T^{n0} does not have the same meaning as T^{0n}. T^{01} is the flux of energy in the x direction, but T^{10} is an entirely different thing; namely, it is the density of the x component of momentum. It's helpful to visualize $T^{\mu\nu}$ as follows:

$$T^{\mu\nu} = \begin{pmatrix} \frac{1}{2}(E^2 + B^2) & S_x & S_y & S_z \\ S_x & -\sigma_{xx} & -\sigma_{xy} & -\sigma_{xz} \\ S_y & -\sigma_{yx} & -\sigma_{yy} & -\sigma_{yx} \\ S_z & -\sigma_{zx} & -\sigma_{zy} & -\sigma_{zz} \end{pmatrix},$$

where (S_x, S_y, S_z) are components of the Poynting vector. In this form, it's easy to see how the mixed space-time components (the top row and leftmost column) differ from the space-space components (the 3×3 submatrix in the lower right). As we noted earlier, σ_{mn} are the components of a tensor called the *electromagnetic stress tensor*, which we have not discussed in detail.

Exercise 11.1: Show that T^{0n} is the Poynting vector.

Exercise 11.2: Calculate T^{11} and T^{12} in terms of the field components (E_x, E_y, E_z) and (B_x, B_y, B_z).

Suppose we restored the factors of the speed of light c that we previously set to 1. The easiest way to do that is by dimensional analysis. We would find that the energy flux and momentum density differ by a factor of c^2. The dimensionally correct identification is that the momentum density is $\vec{E} \times \vec{B}$, with no factor of c. Evidently we have made a discovery, and it's one that has been confirmed by experiment:

For an electromagnetic wave, the density of momentum is equal to the flux of energy divided by the speed of light squared. Both are proportional to the Poynting vector.

Now we can see why on the one hand sunlight warms us when it's absorbed, but on the other hand it exerts such a feeble force. The reason is that the energy density is c^2 times the momentum density, and c^2 is a very big number. As an exercise, you can calculate the force on a solar sail a million square meters in area, at Earth distance from the Sun. The result is tiny, about 8 newtons or roughly 2 pounds. On the other hand, if the same sail absorbed (rather than reflected) the sunlight that fell on it, the power absorbed would be about a million kilowatts.

The density and current of electric charge play a central role in electrodynamics. They appear in Maxwell's equations as the sources of the electromagnetic fields. One may wonder if the energy-momentum tensor plays any similar role. In electrodynamics the answer is no; $T^{\mu\nu}$ does not directly appear in the equations for $\vec{E}$ and $\vec{B}$. It's only in the theory of gravity that energy and momentum take their rightful role as sources, but not sources of electromagnetic fields. The energy-momentum tensor appears in the general theory of relativity as the source of the gravitational field. But that's another subject.

11.6   Bye for Now

Classical field theory is one of the great accomplishments of nineteenth- and twentieth-century physics. It ties together the broad fields of electromagnetism and classical mechanics, using the action principle and special relativity as the glue. It provides a framework for studying *any* field—gravity, for example—from a classical perspective. It's a crucial pre-requisite for the study of quantum field theory and for general relativity (the topic of our next book). We hope we've made the subject understandable and even fun. We're thrilled that you made it all the way to the end.

Some wiseguy once said,

> Outside of a dog, a book is a man's best friend. Inside of a dog it's too dark to read.

If you happen to review our book from either the outside or the inside of a dog—or the boundary of a dog, for that matter—please include some reference to Groucho, however subtle, in your review. We will deem that to be sufficient proof that you actually read the book.

As our friend Hermann might have said, "Time is *up!*" See you in general relativity.

Farmer Lenny, out standing in his ϕield.

A pair o' ducks dances to Art's Φiddle.

Appendix A

Magnetic Monopoles: Lenny Fools Art

"Hey, Art, let me show you something. Here, take a look at this."

"Holy moly, Lenny, I think you found a magnetic monopole. But hold on! Didn't you tell me that monopoles are impossible? Come on, come clean; you've got something up your sleeve and I think I know what it is. A solenoid, right? Ha ha, good trick."

"Nope, Art, not a trick. It's a monopole all right, and no strings attached."

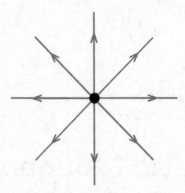

Figure A.1: Electric Monopole. It's just a positive charge.

What would a magnetic monopole be, if there were such things? First, what does the term *monopole* mean? A monopole is simply an isolated charge. The term *electric monopole*, if it were commonly used, would simply mean an electrically charged particle, along with its electric Coulomb field. The convention (see Fig. A.1) is that the electric field points outward for a positive charge like the proton, and inward for a negative charge like the electron. We say that the charge is the source of the electric field.

A magnetic monopole would be exactly the same as an electric monopole, except that it would be surrounded by a magnetic field. Magnetic monopoles would also come in two kinds, those for which the field points outward, and those for which it points inward.

A sailor of old, who might have had a magnetic monopole, would have classified it as either a north or a south magnetic monopole, depending on whether it was attracted to the Earth's north or south pole. From a mathematical point of view it's better to call the two types of magnetic monopoles positive and negative, depending on whether the magnetic field points out or in. If such objects exist in nature, we would call them sources of the magnetic field.

Electric and magnetic fields are mathematically similar, but there is one crucial difference. Electric fields do have sources: electric charges. But according to the standard theory of electromagnetism, magnetic fields do not. The reason is captured mathematically in the two Maxwell equations

$$\vec{\nabla} \cdot \vec{E} = \rho \qquad \text{(A.1)}$$

and

$$\vec{\nabla} \cdot \vec{B} = 0. \qquad \text{(A.2)}$$

The first equation says that electric charge is the source of electric field. If we solve it for a point source we get the good old Coulomb field for an electric monopole.

The second equation tells us that the magnetic field has no sources, and for that reason there is no such thing as a magnetic monopole. Nevertheless, despite this compelling argument, magnetic monopoles not only are possible, but they are an almost universal feature of modern theories of elementary particles. How can this be?

One possible answer is to just change the second equation by putting a magnetic source on the right side. Calling the density of magnetic charge σ, we might replace Eq. A.2 with

$$\vec{\nabla} \cdot \vec{B} = \sigma. \qquad \text{(A.3)}$$

Solving this equation for a point magnetic source would result in a a magnetic Coulomb field that's the exact analog of the electric Coulomb field,

$$B = \frac{\mu}{4\pi r^2}. \qquad \text{(A.4)}$$

The constant μ would be the magnetic charge of the magnetic monopole. By analogy with electrostatic forces, two magnetic monopoles would be expected to experience magnetic Coulomb forces between them; the only difference would be that the product of the electric charges would be replaced with the product of the magnetic charges.

But it's not so simple. We really don't have the option of tinkering with $\nabla \cdot B = 0$. The underlying framework for electromagnetism is not the Maxwell equations, but rather the action principle, the principle of gauge invariance, and the vector potential. The magnetic field is a derived concept defined by the equation

$$\vec{B} = \vec{\nabla} \times \vec{A}. \qquad (A.5)$$

At this point it might be a good idea to go back to Lecture 8 and review the arguments that led us to Eq. A.5. What does this equation have to do with magnetic monopoles? The answer is a mathematical identity, one that we have used several times: *The divergence of a curl is always zero.*

In other words, any vector field such as $\vec{B}$ that's defined as the curl of another field—the vector potential $\vec{A}$ in this case—automatically has zero divergence. Thus $\vec{\nabla} \cdot \vec{B} = 0$ appears to be an unavoidable consequence of the very definition of $\vec{B}$.

Nonetheless, most theoretical physicists are firmly persuaded that monopoles can, and probably do, exist. The argument goes back to Paul Dirac, who in 1931 explained how one could "fake" a monopole. In fact, the fake would be so convincing that it would be impossible to distinguish from the real thing.

Start with an ordinary bar magnet (Fig. A.2) or even better, a electromagnet or solenoid. A solenoid (Fig. A.3) is a cylinder wrapped by wire with current going through it. The current creates a magnetic field, which is like the field of the bar magnet. The solenoid has the advantage that we can vary the strength of the magnet by varying the current through the wire.

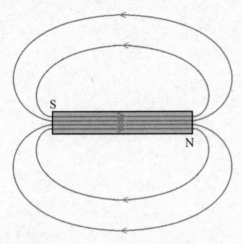

Figure A.2: A Bar Magnet with North and South Poles.

Every magnet, including the solenoid, has a north and south pole that we can call positive and negative. Magnetic field comes out of the positive pole and returns through the negative pole. If you ignore the magnet in between, it looks very much like a pair of magnetic monopoles, one positive and one negative. But of course you can't ignore the solenoid. The magnetic field doesn't end at the poles; it passes through the solenoid so that the lines of flux don't terminate; they form continuous loops. The divergence of the magnetic field is zero even though it looks like there is a pair of sources.

Now let's stretch out the bar magnet or solenoid, making it very long and thin. At the same time let's remove the south (negative) pole to such a far distance that it might as well be infinitely far.

The remaining north pole looks like an isolated positive magnetic charge (Fig. A.4). If one had several such simulated monopoles they would interact much like actual monopoles, exerting magnetic Coulomb forces on each other. But of

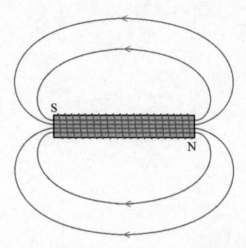

Figure A.3: A Solenoid or Electromagnet. The magnetic field is generated by a current through the wire wound around the cylindrical core.

course the long solenoid is unavoidably attached to the fake monopole, and the magnetic flux passes through it.

We could go further and imagine the solenoid to be flexible (call it a Dirac string), as illustrated in Fig. A.5. As long as the flux goes through the string and comes out at the other pole, Maxwell equation $\vec{\nabla} \cdot \vec{B} = 0$ would be satisfied.

Finally, we could make the flexible string so thin that it's invisible (Fig. A.6). You might think that the solenoid would be easily detectable, so that the monopole could be easily unmasked as a fake. But suppose that it were so thin that any charged particle moving in its vicinity would have a negligible chance of hitting it and experiencing the magnetic field inside the string. If it were thin enough, the string would pass between the atoms of any material, leaving it unaffected.

It's not practical to make solenoids that are so thin that they pass through all matter, but it's the thought experiment that's important. It shows that magnetic monopoles, or at

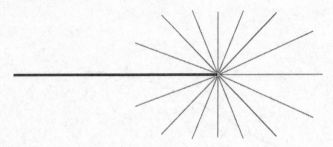

Figure A.4: Elongated Solenoid.

least convincing simulations of them, must be mathematically possible. Moreover, by varying the current in the solenoid, it is possible to make the monopoles at the ends of the solenoid have any magnetic charge.

If physics were classical—not quantum mechanical—this argument would be correct: monopoles of any magnetic charge would be possible. But Dirac realized that quantum mechanics introduces a subtle new element. Quantum mechanically, an infinitely thin solenoid would not generally be undetectable. It would affect the motion of charged particles in a subtle way, even if they never get close to the string. In order to explain why this is so, we do need to use a bit of quantum mechanics, but I'll keep it simple.

Let's imagine our thin solenoid is extremely long, stretching across all space, and that we are somewhere near the string but far away from either end. We want to determine the effect on an atom whose nucleus lies near the solenoid and whose electrons orbit around the solenoid. Classically, there would be no effect because the electrons don't pass through the magnetic field.

Instead of a real atom, which is a bit complicated, we can use a simplified model—a circular ring of radius r that an electron slides along (Fig. A.7). If the solenoid passes through the center of the ring, the electron will orbit around the

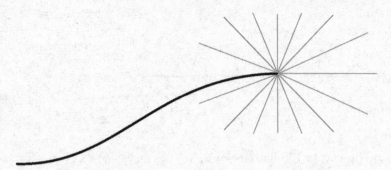

Figure A.5: Dirac String: A Thin and Flexible Solenoid.

Figure A.6: Limiting Case of a Dirac String. The string could be so thin as to be unobservable.

solenoid (if the electron has any angular momentum). To begin with, suppose that there is no current in the solenoid so that magnetic field threading the string is zero.[1] Suppose the velocity of the electron on the ring is v. Its momentum p and angular momentum L are

$$p = mv \qquad\qquad (A.6)$$

and

$$L = mvr. \qquad\qquad (A.7)$$

---

[1] This is where the variable field of the solenoid is helpful.

Figure A.7: Ring Surrounding a Thin Stringlike Solenoid. A charged electron slides along the ring. By varying the current in the solenoid we can vary the magnetic field threading the string.

Finally, the energy ϵ of the electron is

$$\epsilon = \frac{1}{2}mv^2, \qquad (A.8)$$

which we may express as a function of the angular momentum L,

$$\epsilon = \frac{L^2}{2mr^2}. \qquad (A.9)$$

Now for the introduction of quantum mechanics. All we need is one basic fact, discovered by Niels Bohr in 1913. It was Bohr who first realized that angular momentum comes in discrete quanta. This is true for an atom, and just as true for an electron moving on a circular ring. Bohr's quantization condition was that the orbital angular momentum of the electron must be an integer multiple of Planck's constant $\hbar$. Calling L the orbital angular momentum, Bohr wrote

$$L = n\hbar, \qquad (A.10)$$

where n can be any integer, positive, negative, or zero, but nothing in between. It follows that the energy levels of the electron moving on the ring are discrete and have the values

$$\epsilon_n = n^2 \frac{\hbar^2}{2mr^2}. \tag{A.11}$$

Thus far, we have assumed that there is no magnetic field threading the string, but our real interest is the effect on the electron of the string with magnetic flux ϕ running through it. So let's ramp up the current and create a magnetic flux through the string. The magnetic flux starts at zero; over a time interval Δt it grows to its final value ϕ.

One might think that turning on the magnetic field has no effect on the electron since the electron is not where the magnetic field is located. But that's wrong. The reason is Faraday's law: A changing magnetic field creates an electric field. It's really just the Maxwell equation

$$\vec{\nabla} \times \vec{E} = -\frac{\partial \vec{B}}{\partial t}. \tag{A.12}$$

This equation says that the increasing magnetic flux induces an electric field that surrounds the string and exerts a force on the electron. The force exerts a torque and accelerates the electron's angular motion, thereby changing its angular momentum (Fig. A.8).

If the flux through the string is $\phi(t)$, then Eq. 9.18 tells us that the EMF is

$$EMF = -\frac{d\phi}{dt}.$$

Because EMF represents the energy needed to push a unit charge once around the entire loop, the electric field at the ring has the opposite sign and is "spread out" along the

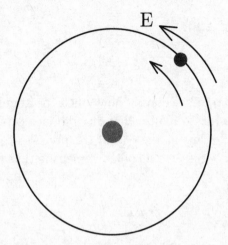

Figure A.8: Another view of the solenoid, ring, and charge.

length of the ring. In other words,

$$E = \frac{\dot{\phi}}{2\pi r},$$ (A.13)

and the torque (force times r) is

$$T = \frac{q\dot{\phi}}{2\pi}.$$ (A.14)

Exercise A.1: Derive Eq. A.13, based on Eq. 9.18. Hint: The derivation follows the same logic as the derivation of Eq. 9.22 in Section 9.2.5.

A torque will change angular momentum in the same way that a force changes momentum. In fact, the change in angular momentum due to a torque T applied for a time Δt is

$$\Delta L = T\Delta t,$$ (A.15)

or, using Eq. A.14,

$$\Delta L = \frac{q\dot{\phi}}{2\pi}\Delta t. \tag{A.16}$$

The final step in calculating how much the angular momentum changes, is to realize that the product $\dot{\phi}\Delta t$ is just the final flux ϕ.[2] Thus at the end of the process of ramping up the flux, the electron's angular momentum has changed by the amount

$$\Delta L = \frac{q\phi}{2\pi}. \tag{A.17}$$

By this time, the flux through the solenoid is no longer changing and therefore there is no longer an electric field. But two things have changed: First, there is now a magnetic flux ϕ through the solenoid. Second, the angular momentum of the electron has shifted by $\frac{q\phi}{2\pi}$. In other words, the new value of L is

$$L = n\hbar + \frac{q\phi}{2\pi}, \tag{A.18}$$

which is not necessarily an integer multiple of $\hbar$. This in turn shifts the possible set of energy levels of the electron, whether the electron lives on a ring or in an atom. A change in the possible energies of an atom would be easy to observe in the spectral lines emitted by the atom. But this happens only for electrons that orbit the string. As a consequence, one could locate such a string by moving an atom around and measuring its energy levels. This would

---

[2] Technically, it's the *change* in flux over time period Δt, but by assumption the flux starts out at zero. Therefore these two quantities are the same.

put the kibosh on Lenny's trick to fool Art with his fake monopole.

But there is an exception. Going back to Eq. A.18, suppose that the shift of angular momentum just happened to be equal to an integer multiple n' of Planck's constant. In other words, suppose

$$\frac{q\phi}{2\pi} = n'\hbar.$$ (A.19)

In that case the possible values of the angular momentum (and the energy levels) would be no different than they were when the flux was zero, namely some integer multiple of Planck's constant. Art would not be able to tell that the string was present by its effect on atoms or anything else. This only happens for certain quantized values of the flux,

$$\phi = \frac{2\pi n\hbar}{q}.$$ (A.20)

Now let's return to one end of the string, say the positive end. The magnetic flux threading the string will spread out and mimic the field of a monopole just as in Fig. A.6. The charge of the monopole μ is just the amount of flux that spills out of the end of the string; in other words it is ϕ. If it is quantized as in Eq. A.20, then the string will be invisible, even quantum mechanically.

Putting it all together, Lenny can indeed fool Art with his "fake" magnetic monopole, but only if the charge of the monopole is related to the electron charge q by

$$\mu q = 2\pi\hbar n.$$ (A.21)

The point, of course, is not just that someone can be fooled by a fake monopole; for all intents and purposes, real monopoles can exist, but only if their charges satisfy

Eq. A.21. The argument may seem contrived, but modern quantum field theory has convinced physicists that it's correct.

But this does leave the question of why no magnetic monopole has ever been seen. Why are they not as abundant as electrons? The answer provided by quantum field theory is that monopoles are very heavy, much too massive to be produced in particle collisions, even those taking place in the most powerful particle accelerators. If current estimates are correct, no accelerator we are ever likely to build would create collisions with enough energy to produce a magnetic monopole. So do they have any effect on observable physics?

Dirac noticed something else: The effect of a single monopole in the universe, or even just the possibility of a monopole, has a profound implication. Suppose that there existed an electrically charged particle in nature whose charge was not an integer multiple of the electron charge. Let's call the electric charge of the new particle Q. A monopole that satisfies Eq. A.21, with q being the electron charge, might not satisfy the equation if q were replaced by Q. In that case, Art could expose Lenny's fraudulent monopole with an experiment using the new particle. For this reason, Dirac argued that the existence of even a single magnetic monopole in the universe, or even the possibility of such a monopole, requires that all particles have electric charges that are integer multiples of a single basic unit of charge, the electron charge. If a particle with charge $\sqrt{2}$ times the electron charge, or any other irrational multiple, were to be discovered, it would mean that monopoles could not exist.

Is it true that every charge in nature is an integer multiple of the electron charge? As far as we know, it is. There are electrically neutral particles like the neutron and neutrino, for which the integer is zero; protons and positrons whose charges are -1 times the electron charge; and many others. So far, every particle ever discovered—including composites like nuclei, atomic ions, exotics like the Higgs boson, and all others—has a charge that's an integer multiple of the electron charge.[3]

---

[3] You may think that quarks provide a counterexample to this argument, but they don't. It's true that quarks carry charges of $\pm\frac{1}{3}$ or $\pm\frac{2}{3}$ of the electron charge. However, they always occur in combinations that yield an integer multiple of the electron charge.

or

$$\vec{\nabla} \cdot \vec{A} = \frac{\partial A_x}{\partial x} + \frac{\partial A_y}{\partial y} + \frac{\partial A_z}{\partial z}.$$

The divergence of a field at a specific location indicates the tendency of that field to spread out from that point. A positive divergence means the field spreads out from this location. A negative divergence means the opposite—the field tends to converge toward that location.

B.4 Curl

The curl indicates the tendency of a vector field to rotate or circulate. If the curl is zero at some location, the field at that location is irrotational. The curl of $\vec{A}$, written $\vec{\nabla} \times \vec{A}$, is itself a vector field. It is defined to be

$$\vec{\nabla} \times \vec{A} = (\partial_y A_z - \partial_z A_y)\hat{\mathbf{i}} + (\partial_z A_x - \partial_x A_z)\hat{\mathbf{j}} + (\partial_x A_y - \partial_y A_x)\hat{\mathbf{k}}.$$

Its x, y, and z components are

$$(\vec{\nabla} \times \vec{A})_x = \partial_y A_z - \partial_z A_y$$

$$(\vec{\nabla} \times \vec{A})_y = \partial_z A_x - \partial_x A_z$$

$$(\vec{\nabla} \times \vec{A})_z = \partial_x A_y - \partial_y A_x,$$

or

$$(\vec{\nabla} \times \vec{A})_x = \frac{\partial A_z}{\partial y} - \frac{\partial A_y}{\partial z}$$

$$(\vec{\nabla} \times \vec{A})_y = \frac{\partial A_x}{\partial z} - \frac{\partial A_z}{\partial x}$$

$$(\vec{\nabla} \times \vec{A})_z = \frac{\partial A_y}{\partial x} - \frac{\partial A_x}{\partial y}.$$

For convenience, we'll rewrite the preceding equations using numerical indices:

$$(\vec{\nabla} \times \vec{A})_1 = \frac{\partial A_3}{\partial X^2} - \frac{\partial A_2}{\partial X^3}$$

$$(\vec{\nabla} \times \vec{A})_2 = \frac{\partial A_1}{\partial X^3} - \frac{\partial A_3}{\partial X^1}$$

$$(\vec{\nabla} \times \vec{A})_3 = \frac{\partial A_2}{\partial X^1} - \frac{\partial A_1}{\partial X^2}.$$

The curl operator has the same algebraic form as the vector cross product, which we summarize here for easy reference. The components of $\vec{U} \times \vec{V}$ are

$$(\vec{U} \times \vec{V})_x = U_y V_z - U_z V_y$$

$$(\vec{U} \times \vec{V})_y = U_z V_x - U_x V_z$$

$$(\vec{U} \times \vec{V})_z = U_x V_y - U_y V_x.$$

Using index notation, this becomes

$$(\vec{U} \times \vec{V})_1 = U_2 V_3 - U_3 V_2$$

$$(\vec{U} \times \vec{V})_2 = U_3 V_1 - U_1 V_3$$

$$(\vec{U} \times \vec{V})_3 = U_1 V_2 - U_2 V_1.$$

B.5 Laplacian

The Laplacian is the divergence of the gradient. It operates on a twice-differentiable scalar function S, and results in a scalar. In symbols, it's defined as

$$\nabla^2 = \vec{\nabla} \cdot \vec{\nabla}.$$

Referring back to Eq. B.1, this becomes

$$\nabla^2 = \left(\frac{\partial}{\partial x}\hat{\mathbf{i}} + \frac{\partial}{\partial y}\hat{\mathbf{j}} + \frac{\partial}{\partial z}\hat{\mathbf{k}}\right) \cdot \left(\frac{\partial}{\partial x}\hat{\mathbf{i}} + \frac{\partial}{\partial y}\hat{\mathbf{j}} + \frac{\partial}{\partial z}\hat{\mathbf{k}}\right)$$

$$\nabla^2 = \frac{\partial^2}{\partial x^2} + \frac{\partial^2}{\partial y^2} + \frac{\partial^2}{\partial z^2}.$$

If we apply ∇^2 to a scalar function S, we get

$$\nabla^2 S = \frac{\partial^2 S}{\partial x^2} + \frac{\partial^2 S}{\partial y^2} + \frac{\partial^2 S}{\partial z^2}.$$

The value of $\nabla^2 S$ at a particular point tells you how the value of S at that point compares to the average value of S at nearby surrounding points. If $\nabla^2 S > 0$ at point p, then the value of S at point p is less than the average value of S at nearby surrounding points.

We usually write the ∇^2 operator *without* an arrow on top because it operates on a scalar and produces another scalar. There is also a vector version of the Laplacian operator, written *with* an arrow, whose components are

$$\vec{\nabla}^2 \vec{A} = \left(\nabla^2 A_x, \nabla^2 A_y, \nabla^2 A_z\right).$$

Index

Leonard Susskind is the Felix Bloch Professor in Theoretical Physics at Stanford University. He is the author of *Quantum Mechanics* (with Art Friedman) and *The Theoretical Minimum* (with George Hrabovsky), among other books. He lives in Palo Alto, California.

Art Friedman is a data consultant and the author of *Quantum Mechanics* (with Leonard Susskind). A lifelong student of physics, he lives in Murphys, California.